About the author

After graduating from Trinity College, Cambridge with a degree in Classics, Martin Fone started his working career as an audit assistant. However, he soon found the world of bean-counting too racy for his taste and retreated to the calmer pastures of the insurance industry. He had a successful business career, during the course of which he co-authored two books on public sector risk management, which were adopted by the Institute of Risk Management as their standard text books.

Since retiring, Martin has had the opportunity to develop his interests, mainly reading, writing and thinking or, as his wife puts it, "locking himself away in his office for a few hours a day". In particular, he has been blogging and writing in his tongue-in-cheek, irreverent style about the quirks, idiocies and idiosyncrasies of life, both modern and ancient.

This is the fourth book he has written since leaving the insurance industry behind, following on from *Fifty Clever Bastards, Fifty Curious Questions* and *Fifty Scams and Hoaxes*, all of which, he says, are still available from all good book retailers and high-class charity shops. Martin also contributes to *Country Life Online*.

THE

FICKLE

FINGER

AN INVENTOR'S LOT

Martin Fone

Matador
9 Priory Business Park,
Wistow Road, Kibworth Beauchamp,
Leicestershire. LE8 0RX
Tel: 0116 279 2299
Email: books@troubador.co.uk
Web: www.troubador.co.uk/matador
Twitter: @matadorbooks

ISBN 978 1838593 117

British Library Cataloguing in Publication Data.
A catalogue record for this book is available from the British Library.

Typeset in 11pt Bembo by Troubador Publishing Ltd, Leicester, UK

Matador is an imprint of Troubador Publishing Ltd

This book is dedicated to my parents, Ray and Brenda.
I am also eternally indebted to my wonderful wife, Jenny,
whose love and support made this book possible.

CONTENTS

PART THREE
IMITATION AND NEGLECTED SPARKS   85

PART FOUR
PATENTLY UNFAIR   125

INTRODUCTION

AMBITION DRIVES YOU on, ability certainly helps, but the fickle finger of fate and luck are great things
– Fergus Henderson

Ah, meritocracy! The term was coined by Michael Young in his dystopian account of a world in which the gifted, the smart, the energetic, the ambitious and the ruthless were selected to fulfil their rightful roles in society, *The Rise of the Meritocracy* (1958). In common parlance, Young's intended negative connotations have all but disappeared, the word now conjuring up the image of a society where one's advancement (or otherwise) is entirely dependent upon one's own merits. As a young man, I was quite taken by the idea.

After all, in my particular case, the circumstantial evidence that success, however defined, could be earned solely through one's own efforts and talents was compelling.

My parents had pulled themselves up by their boot straps from a working-class background to enjoy an, ultimately, comfortable middle-class lifestyle. As a state schoolboy, I had won a place in one of the world's top universities.

As I became more worldly-wise, though, I quickly realised that there were other, bigger, forces at play. On countless occasions I read, heard or experienced for myself people attributing their success to getting that lucky break or being in the right place at the right time. Does success always have to involve a healthy slice of luck and a benevolent fickle finger of fate waving in your direction, as Fergus Henderson suggests?

The fickle finger of fate, a wonderfully alliterative phrase, conveys the sense of man's powerlessness against a combination of circumstances that either spur him on to greatness or conspire against him and rob him of his just desserts. I am not a determinist, I do believe that we have the freedom to make our own choices, but isn't it strange how some people seem to have more good fortune than others while some have more than their fair share of bad luck?

While reading about the success stories of others can be instructive, I'm more of a glass half empty sort of person. I'm much more interested in why some people were not as successful as, at first blush, they should have been. In this, a light-hearted investigation into the greater powers which play a part in determining success or failure, I have chosen the world of the inventor as my field of reference.

The ability to work through a problem and find a better solution, to spot a way to improve our daily lot, to enable Homo sapiens to escape the restrictions imposed on him by bipedalism, these qualities all fascinate me. I have great admiration for those amongst us who have an inventive streak, who are be able to think outside of the box and come up with an idea or a design which has a transformational effect on the way we live our daily lives.

It is tempting to think that once the grey cells have whirred and a prototype has been made, the idea proven and launched on the unsuspecting public, all the inventor then has to do is sit back and watch the money roll in. In many instances, that is the case.

However, as I investigated the stories of some of the lesser known inventors, I began to realise that there was a bigger theme to explore, the interplay between the way in which the world operates and success. This book explores, through a series of fifty vignettes, some of the obstacles that life, or perhaps fate, has thrown in the inventor's way to deprive them of, or at least diminish, the financial gains and glory that their brainwave merited.

Some are rooted in the mores and prejudices of society at the time, some are related to inadequate or non-existent legal protections, some are to do with inferior PR, and others are the result of some catastrophic blunder or poor decision-making.

Assuming that their invention has not killed them or that their demonstration of the virtues of their invention has not been so ostentatiously disastrous as to set the public against it (Nos. 1 to 3), or that they don't have bigger fish to fry (Nos. 4 to 7), or that they are not imbued with the spirit of philanthropy or are just simply oblivious to the importance of what they have done (Nos. 8 to 13), then there are more difficult hurdles to overcome.

The inventive streak is not restricted to white males. Women and people of colour, whether slaves or free, also had brains and were able to think through problems, gain a greater understanding of the universe in which we live or improve the way we do things. In order to get their ideas to a wider audience they had to battle institutionalised sexism (Nos. 14 to 19) and racism (Nos. 20 to 25). Indeed, as we shall see when we meet Benjamin T Montgomery (No. 19), at one time slaves were not able to assert their intellectual property rights. They were not deemed to be citizens of the United States and, consequently, were unable to swear an oath to the effect that the invention was their own.

Others simply had their claims to an invention usurped by others with what we would call today a more powerful and effective PR machine (Nos. 26 to 33). Many an invention that we attribute to one historical figure simply wasn't theirs. They may have improved upon the original concept, but the original spark of genius belongs to others who have long been relegated to the footnotes of history.

Some inventors were so ahead of their time and their ideas were so counter-intuitive to the prevailing wisdom of the time that they were, at best, ignored or publicly ridiculed (Nos. 34 to 38). They were later proven to be right but the fight to get their ideas was exhausting and often detrimental to their health and sanity.

Then we come to patents.

English monarchs were able to offer "letters patent" or *literae patentes*, which gave the bearer a monopoly to exploit whatever was the subject of the grant. However, the system fell into some disrepute, when patents were extended to common or garden goods, such as salt. After a public outcry, King James the First abolished all existing monopolies and, in 1624, passed the Statute of Monopolies. The new law restricted the issuance of patents to inventors or introducers of "projects of new invention" and, then, only for a limited period of time. The patent system, as we know it, was born.

Other countries, keen to foster and protect the inventive spirit of their denizens, followed suit. There clearly wasn't a patent on that idea. The US Congress passed the snappily entitled "An Act to promote the progress of useful Arts" on April 10, 1790, the first patent being granted, on 31st July of that year, to Samuel Hopkins for a method of producing potash.

Abraham Lincoln, we will meet him a few times in this book, was a fervent advocate of the system, calling it that which added "the fuel of interest to the fire of genius" and Mark Twain espoused the idea that a country without a patent system was like a crab, only able to move "sideways and backwards."

The idea behind a patent was simple enough. It was designed to give an inventor sufficient time to exploit commercially their idea without granting them an everlasting monopoly on the idea. That was fair enough but in the hands of unscrupulous or overly sensitive corporations, it is a system which can stifle the enterprise and initiative of entrepreneurs working in similar areas.

For the little man, on whom I focus in this book, the patent system exposed some limitations, which made life difficult and throw some doubt upon the overall efficacy of the system. In the first place, the inventor had to make sure that their patent was renewed and in place throughout the period for which it is granted. We will come across some individuals who, through either lack of cash or through neglect, omitted to renew their patent or having been told there was no market for their idea, were astonished to find that it was taken up with some gusto by larger corporations once the time limit had expired (Nos. 39 to 42).

Then there was the problem of patent infringements or subtle changes in design by competitors which rendered the patent held by the inventor ineffective. The inventor was left with little choice other than to accept this incursion on to their turf or to engage in lengthy and expensive legal action to reassert the claim. Often, the wearying process took its toll on their resources and health, raising the question of whether the patent system was really weighted in favour of those rich in resources like corporations who could defend their invention against others or frustrate inventors seeking to exploit their ideas rather than offering the protection originally envisaged and needed by the struggling inventor (Nos. 43 to 48).

Over time, a secondary market developed for patents which could now be perceived as a tradeable asset. This meant that an inventor could sell their patent or a percentage of it, swapping uncertain future revenue for immediate cash, that is, if the purchaser was able to pay up or the inventor had placed a sensible valuation on their patent. For some, who were strapped for cash, selling their patent was too much of a temptation to pass up. But it did not always end well (Nos. 49 to 50).

As the narrative develops, the attentive reader will note some other trends that recur. Several stories tell of the struggles that the inventor had with larger organisations to obtain their just rewards or the recognition that their endeavours deserved or simply just to get on with running their business and exploiting their invention. The establishment or those with turf to protect do not take kindly to the incursions of smart alec inventors. Often, the struggle involved many years fighting for justice, a wearying process which took its toll on the inventor's resources and health, even driving some to commit suicide.

Another theme running through the book is the role of the wife. For every Mrs Rontgen (No. 8), who was mortified by the sight of an X-ray of her hand, there is a Mrs Farnsworth (No. 44), who continued her husband's fight for justice after his death. There may just be something in the adage that behind every great man there is a great woman.

To succeed an inventor needs a healthy dose of luck and determination but the fickle finger of fate can play cruel tricks, as we shall see.

Part One

Catastrophes, Fickle Fame and Philanthropy

THE CATASTROPHIC ERROR

ASK YOURSELF THIS, "Is this a blip or is it a catastrophe?" – Donald J Trump, tweet, December 30, 2013, 8.27 am.

There is a fine line between comedy and tragedy. One person's misfortune can often be a source of amusement to another. Take slipping on a banana skin. The unfortunate victim will find that their dignity and, perhaps, their posterior has been bruised but when they get back up onto their feet, somewhat gingerly, they will perceive a lack of sympathy from any onlookers who will be barely able to contain their mirth.

In the grand scheme of things, their embarrassment can only be described as a blip, at the very most, and normal life swiftly reverts to the norm. The Trumpian scale of blip and catastrophe needs to be reserved for much weightier events, ones that can have a much deeper transformational effect on the development of man's knowledge of the environment in which he lives and the way he can shape it to his will.

When I first started my investigations into the inventor's lot, I quickly found that if I was not too careful, I would easily be accused of serving up a feast of schadenfreude, inviting you, dear readers, to relish the discomfiture and misfortunes of those who attempted, with mixed success, to push out the boundaries of human accomplishment and knowledge.

So, I have tempered my enthusiasm. After all, the nineteenth century German philosopher, Arthur Schopenhauer, opined in *On Human Nature* that "to savour schadenfreude is devilish". However, it would be remiss of me to ignore this aspect of their lot completely. I must have a devilish streak in me. Here are the stories of three individuals who in very different ways contributed to their own downfall, two fatally, and the other delaying our enjoyment of the fruits of their little grey cells for nigh on a century.

Whether they are blips or catastrophes, I will leave you to decide.

I. Dr. Louis Slotin (1910 – 1946)

We've all done it, I'm sure, moaned about the red tape of bureaucracy and "'elf and safety", which hinders us from getting on with what we are trying to do.

Occasionally, there are good reasons why a bit of safety awareness wouldn't come amiss, as this cautionary tale involving Canadian scientist, Louis Slotin, amply illustrates.

The Manhattan Project was a research project undertaken during the Second World War which, ultimately, saw the development of the world's first nuclear bomb. Slotin was one of the scientists involved at the time and earned a reputation as being one of the pre-eminent assemblers of nuclear warheads.

Following the destruction of Horoshima and Nagasaki and the conclusion of the war, Slotin continued to experiment with nuclear fission. His particular sphere of interest was measuring the beginnings of the fission reaction, by bringing two semi-spherical pieces of radioactive material into close proximity. Of course, if the two actually touched, there would be an almighty explosion and so a degree of precision, as well as a steady hand, was called for.

For some people, playing your part in developing something that could fry large portions of the world's population is not enough. Slotin was a bit of a character who liked to spice up his life. That may be the reason why he eschewed any of the fancy-dan safety equipment available and relied upon a humble screwdriver to keep the two hemispheres apart.

On May 21, 1946, Slotin was training a colleague, the aptly named Alvin Graves, at the Omega Laboratory and for his pièce de resistance, a small crowd of his colleagues assembled to watch his performance. Unfortunately, at the critical moment at around three o'clock in the afternoon, the screwdriver slipped and the two pieces of radioactive material made contact. The official report into the incident recorded that, "the blue flash was clearly visible in the room although it (the room) was well illuminated from the windows and possibly the overhead lights. . . . The total duration of the flash could not have been more than a few tenths of a second."

Showing a remarkable presence of mind, Slotin pushed the top hemisphere of plutonium off with his bare hands, thus ending the reaction.

It was calculated that Slotin's screwdriver slip had set off about three quadrillion fission reactions, it sounds a lot but the bang, in fact, was about a million times smaller than the first atomic bombs. The blue flash was caused by the high-energy photons emitted when the electrons in the air settled down after their agitation.

But the damage was done.

Slotin immediately complained of a burning sensation in his left hand and a sour taste in his mouth. He was rushed into a car and taken to hospital, but during the journey started to vomit, a symptom of severe radiation poisoning.

Slotin said to his colleagues, 'You'll be OK, but I think I'm done for.'

He was not wrong, dying nine days later of radiation exposure.

He was commended for his actions in a citation read to him before meeting his maker. "Dr Slotin's quick reaction at the immediate risk of his own life prevented a more serious development of the experiment which would certainly have resulted in the death of the seven men working with him, as well as serious injury to others in the vicinity." It was a rather optimistic assessment; within two years of the incident, two of his colleagues had died of radiation sickness.

Clearly, Slotin's approach to the experiment had been cavalier. After all, there had been an incident a few months earlier when Harry Daghlian dropped a brick of tungsten carbide on to a plutonium mass, bathing him in radiation. He died a month later from radiation sickness.

Next time you moan about safety regulations, give a thought to Louis Slotin.

2. Dr Sabin Arnold von Sochocky (1883 – 1928)

Born in the Ukraine, von Sochocky migrated to the United States and became an American citizen. As well as being the victim of his own cleverness, he was the cause of the painful demise of many others, as we shall see.

One of the problems with human sight, compared with many other creatures inhabiting the Earth, is that it is pretty ineffective in the dark. We need artificial aid to see where we are going, what we are doing and what is around us. It is hard for us to comprehend these days but until the late nineteenth century human existence was pretty dark, once the sun went down. Of course, candles made some difference, but the light thrown out by them was pretty meagre and the by-products of smoke and smell (from the cheaper tallow candles) were unpleasant.

The introduction of, first, gas lighting and, later, electric lighting was revolutionary and life-changing but was a relatively expensive solution to the inability to see at night.

The race was on to find a cheaper and effective alternative.

The experiments of the husband and wife duo, Marie and Pierre Curie, with radium seemed to offer a solution. Returning to Paris in 1902 with a gift of radium salt crystals from the Curies, William Hammer started a series of experiments in which he combined them with glue and zinc sulphide to produce an iridescent paint. It could be applied to almost anything and the result was astonishing, you might say hair-raising or jaw-dropping.

Whatever was painted, glowed in the dark.

Sochocky was the first to see the commercial potential of this breakthrough.

He improved its qualities and production methods so that it was capable of being mass-marketed and, in 1915, with some colleagues founded the Radium Luminous Materials Co in Orange, New Jersey.

It later changed its name to the U.S Radium Corporation.

The paint was called Undark, perhaps an unwitting reference to its zombie-like qualities, and whilst the marketing puff was up-front about the presence of radium, it reassuringly stated that the substance was present in "such minute quantities that it is absolutely harmless".

Business boomed and the company had a commission from the US military to produce watches and instrumentation with luminous dials. The factory employed over one hundred workers, mainly young girls on piece-work, receiving the princely sum of 1.5 cents an item.

Since the paint was odourless and colourless, the workers were encouraged to dip their paint brush into the substance, then put the tip into their mouth to form a point (and thereby ingest a minute dose of radium each time) and then apply it to whatever they were painting.

During the early 1920s, the first of a rash of mysterious illness and bone injury amongst the dial painters began to occur. By 1924 the situation was so worrying, nine girls had died by then, that the medical examiner of Essex County began to investigate the factory, its production methods and measure the radioactivity present in the dial painters' bodies. His findings, published in 1925, incontrovertibly linked their bone disease and aplastic anaemias with the radium component in the paint. Inevitably the matter led to litigation as employees, or the relatives of former employees, sought compensation. The first case concerned Grace Fryer and was settled in her favour in 1928.

Nevertheless, the company didn't stop the practice of hand painting dials until 1948.

Sochocky himself succumbed to consequences of radium poisoning in 1928. During his illness his teeth dropped out and his fingers turned black and he was only able to prolong his life by taking frequent holidays at high altitude, a well-known cure for anaemia.

It may be a harsh judgment but there is some justice, after all.

3. John Joseph Merlin (1735 – 1803)

Presentation is everything. Perhaps one of the most catastrophic unveilings of an invention ever took place in London in 1771. The repercussions were such that the device did not see the light of day again for another ninety years.

Born in Huy in Belgium, John Joseph Merlin studied at the Academie des Sciences in Paris where he became well-known for his inventiveness. He moved to London in 1760, initially as the technical adviser to the Spanish Ambassador, but by 1766 was working with James Cox, a jeweller and goldsmith, as what was termed a mechanician.

Merlin went on to use his knowledge of automata and the mechanics of clocks to develop a range of innovative toys and musical instruments, which he patented. In 1783 he opened, in Hanover Square, a mechanical museum where he displayed many of the toys and objects that he had developed.

It was a great success, Madame d'Arblay noting that "Merlin was quite the rage in London where everything was à la Merlin: Merlin chairs, [he had developed a mechanical gouty chair] Merlin pianos, Merlin swings... Merlin fiddles and Merlin mechanical pegs for violins and violoncellos".

Merlin was lionised by the great and the good.

He was good friends with Thomas Gainsborough, who painted a rather splendid portrait of him, possibly in return for one of Merlin's mechanical instruments. He was a regular visitor to the house of the musicologist, Charles Burney.

His daughter, Fanny, wrote that Merlin was "very diverting in conversation…he speaks his opinion upon all subjects and about all persons with the most undisguised freedom". But showing a little Englander attitude that prevailed even then, she noted that "he does not, though a foreigner, want words but he arranges and pronounces them very comically".

Frustration with the limitations that bipedalism imposes on the ability to get from A to B as quickly as possible was something that caused many an inventors' juices to flow, leading on to the development of bicycles, air flight, steam power, submarines, and the like.

Merlin applied his ingenuity to the problem of how to accelerate man's ability to travel. His light bulb moment was to hit upon the ice skate, from which he removed the blade and replaced it with a couple of wheels. Attaching them to the feet he had made, and, naturally, patented the first pair of roller skates.

Merlin was the consummate showman and could not resist the opportunity to demonstrate his roller skates at one of the premier events of the 1771 London season, a soirée at the home of Mrs Cowley's at Carlisle House. For maximum effect, Merlin decided to make his entrance on his roller skates while playing the violin, and why not?

What happened next is to be found in Thomas Busby's *Concert Room and Orchestra Anecdotes* of 1805. "When not having provided the means of retarding his velocity, or commanding its direction he impelled himself against a mirror of more than five hundred pounds value, dashed it to atoms, broke his instrument to pieces and wounded himself most severely".

Merlin's dramatic entrance set back the development of the roller skate by nearly ninety years.

In 1863, James Plimpton, an American, came up with the idea of a skate with four wheels for stability and independent axles, which he called a rocking skate or rocker. So successful was Plimpton's device that roller skating took off. Plimpton's design is still used today.

I wish I had been there to see Merlin's demonstration, though.

TOO FAMOUS BY HALF

FAME IS A fickle food / upon a shifting plate/ whose table once a / guest but not /the second time is set – Emily Dickinson, *Fame is a fickle food* (1702)

Fame is a fickle thing. Very few of us are blessed with even a modicum of fame, living a humdrum existence, noticed only, we hope, by our employers, family and friends. Those of us who do achieve a moment in the spotlight soon find that it is a very transitory experience, lasting little more than the fifteen minutes that Andy Warhol prophesied each of us would attain.

There are some, though, who are blessed enough not only to have a wide range of skills at their disposal but also to excel in more than one field of human endeavour. For us mere mortals, this causes us some mental perturbation.

Perhaps because of the limitations of our brain or, at least, our memory, we find it difficult to associate someone with more than one area of expertise. Our mental shorthand pigeon holes them to the area where they have achieved pre-eminence, putting their other areas of expertise into the shadows, soon to be long forgotten.

Here are four famous individuals who, in their own way, made significant contributions to the advancement of the lot of humans, three of whom could rightly be claimed to have been among the leading figures of their respective centuries. However, they also had other strings to their bow and what has slipped into the mists of time is their inventive streaks.

It is time to redress the balance, methinks.

4. Albert Einstein (1879 – 1955)

If you were to draw up a list of the leading lights of the twentieth century, then Albert Einstein would be right up there. After all, he was a cut above the average in the brainbox stakes, best known for that simple but mind-bogglingly fiendish equation, $E=mc^2$.

But just to demonstrate the relativity of fame, luck and success, did you know that he was involved in the development of a different form of cooling system for a refrigerator?

Here is Albert's tale.

Reading his newspaper over breakfast some time in 1926, Albert came across a tragic story concerning a family in Berlin who had been killed in their sleep. A faulty seal in their fridge had led to an escape of toxic gas, a not uncommon occurrence at the time. What was unusual, though, was the fact that so many in one house had succumbed to this silent form of killer.

Interest piqued, Einstein began to consider whether there was a better and less deadly way of keeping perishable foodstuffs fresh. Enlisting the help of a former student of his, Leo Szilard, he set about solving the problem of the faulty seals.

The root of the problem, they concluded, was the moving parts in the existing fridges, particularly the mechanical compressor. Their revamp would be bold, eliminating the reliance upon mechanical parts.

First up, instead of using a mechanical compressor to manufacture the refrigerant, Einstein and Szilard used a natural gas flame to heat a chemical solution, producing, by way of a chemical reaction, a perfectly serviceable coolant, a technique known nowadays as absorption refrigeration.

The next problem was how to move the coolant around the refrigerator.

The inventive duo decided to develop an electromagnetic pump. In line with their overall design philosophy, there were to be no mechanical parts. Instead, Einstein and Szilard ran an electric current through several separate coils within the fridge, creating an electromagnetic field. The energy within the field was sufficient to pump a liquid metal alloy around the system at such a force that it compressed the refrigerant, as though it was a piston.

Having completed the design, a friend and former student of Einstein's, Albert Korodi, was recruited to build it. Build it he did and it worked a treat, except for one tiny problem. The metal alloy moving through the pump made quite a racket.

Korodi solved the noise problem by adding more electrical coils to the pump and varying the electrical charge that passed through it. In a further modification the radioactive properties of the potassium-sodium alloy used as the liquid metal meant that a specially sealed unit had to be used.

Satisfied with the finished product, Albert applied for a patent which was granted in the States on November 11, 1930. He also obtained patents in 45 other countries.

What went wrong?

Timing is everything. Although the Swedish manufacturers, Electrolux, paid Einstein and Szilard the princely sum of $750 for one of their patents, the Great Depression of 1929 meant that manufacturers didn't have the capital to finance the development of a commercial fridge based on Albert's prototype.

In any event, it was less energy-efficient than other cooling systems around.

What really did for it, though, was the discovery of a non-toxic refrigerant, Freon, in 1930 which eliminated the requirement for non-mechanical compressors.

As a chlorofluorocarbon (CFC), though, the widespread adoption of the cheaper cooling system put a massive hole in the ozone layer but, hey, never mind about that.

So far as the domestic refrigerator was concerned, Einstein's pump was redundant and would have been consigned to the dustbin of history but for one saving grace.

When the world was ready to embrace nuclear technology, the reactors required an efficient cooling system to assist in the production of the radioisotopes.

What system was there better to use than Einstein's?

And so his electromagnetic pump did find a home, just not the one he expected and, of course, he didn't make any money out of it nor was his contribution to refrigerator technology widely recognised.

Still, Einstein may have appreciated the irony of the relativity of fame, success and luck.

5. Abraham Lincoln (1809 – 1865)

For me the pleasures to be derived from boating are outweighed by the hard work involved. As a non-swimmer, I am particularly concerned that, through my clumsiness or incompetence, I will end up in the water.

You could call it one of my hang-ups.

Getting hung up is a serious concern for boating aficionados and describes a scenario where the vessel has become caught up or sits on a sill or ledge, natural or otherwise. Unable to move, the boat becomes unstable, particularly if the water level changes, and, without care, can capsize.

It is not an uncommon problem.

Before gaining the keys to the White House and, ultimately, settling down to watch the Tom Taylor's farce, *Our American Cousin*, at Ford's Theatre in Washington D.C. on April 14, 1865, Abraham Lincoln had a long and varied career.

In his youth, he worked on a flatboat which made its way down from Illinois to New Orleans. Somewhere near New Salem, on the Sangamon River, the boat got hung up on a milldam, started to take on water and began to sink. The only thing for it was to move the cargo around the boat and drain it before the boat capsized. Years later, now a congressman and returning to Illinois, the boat he was travelling on became hung up on some shallows. The captain, with remarkable presence of mind, ordered his crew to put all the moveable items in the ship, such as barrels, boxes and planks of wood, overboard and manoeuvre them underneath. This had the effect of lifting the boat off the shelf and working it free.

These experiences gave Lincoln pause for thought and he set about exercising his little grey cells, he didn't wear a tall hat for nothing, to find a solution to this commonplace boating hazard.

The idea he came up with was a series of bellows of "india-rubber cloth, or other suitable water-proof fabric" attached to the underside of the vessel. In the event that the boat got into difficulties, "by turning the main shaft or shafts in one direction, the buoyant chambers will be forced downwards into the water and at the same time expanded and filled with air", thus raising the boat off the shelf that it was sitting on.

Lincoln even went so far as to create a working model of his invention. His law partner at the time, William H Herndon, noted that Lincoln would bring a model of a boat made out of wood into the office and "while whittling on it would descant on its merits and the revolution it was destined to work in steamboat navigation".

There is a two-foot long, wooden model of Lincoln's design extant in Washington, although whether it was made by him is open to question. Many have noted that his name is misspelled as Abram Lincoln. Would he have been so careless? Some attribute the model to a Springfield mechanic, Walter Davis, who was known to have helped Lincoln, others to an unknown model maker in Washington, who made it his business to help would-be inventors.

By 1849, Lincoln was practising law back in Springfield, after just one term as a congressman. Convinced of the efficacy and practicality of his design, he applied, with the help of a lawyer, Z C Robbins, for a patent for a device for "buoying vessels over shoals". Patent 6469 was granted on May 22, 1849, making Lincoln the first and only US President to have held a patent. So far, at least, no one has trumped him.

The invention, though, never saw the light of day and Lincoln's revolution in steamboat navigation was stillborn.

Some experts, in any event, think that whilst plausible in theory, the invention was unlikely to have been very practical, because of the amount of force that would have been required to inflate the bellows to a size that they would lift the boat off the obstruction.

The design might have been capable of modification but by that time Lincoln's mind and time was occupied on greater matters.

6. Thomas Paine (1737 – 1809)

From the age of around ten, until I went to university, I lived in the beautiful rural county of Shropshire. One of its principal claims to fame is that it is home to the world's first major bridge to be constructed entirely of cast iron, spanning the Severn Gorge at Coalbrookdale.

Abraham Darby's iconic design was a testament to the burgeoning age of industrialisation and word of the bridge, opened in 1779, spread around the world. Its fame gave its rather prosaic name to the small town that grew up around it, Ironbridge.

Revolutionary as the material used to build the bridge was, Darby's iron construction was traditional in design, consisting of five ribs, forming a semicircle, a technique dating back to at least the Roman times. The drawback with a semicircle was that the width dictated the height of the bridge, fine for a steep gorge like the one at Coalbrookdale but creating an irritating hump on wider spans.

The Romans, ingenious to the last, solved this problem by using a sequence of small arches. But this approach caused other problems, not least that there was more work required to secure the footings which, in turn, could alter the flow of the river, as the nineteen arches on London Bridge had done to the Thames.

Now that there was a revolutionary new material with which to construct a bridge, wouldn't it be great if the design was freed from the restrictions imposed by the traditional semicircle methodology?

This is where Thomas Paine sought to make his mark.

One of America's founding fathers, Thomas is best known these days as the author of *The Rights of Man*, published in 1791 and a forthright defence of the French Revolution against the attacks of British politicians such as Edward Burke. But he had other strings to his bow, not least being an ardent pontist, fascinated by the mix of architectural splendour and sheer practicality that makes up a bridge.

Aren't we all?

Intrigued by iron bridges, Paine sought to raise enough money to build a bridge that would span the river Harlem in New York in 1785 and another to cross the Seine in Paris in 1786. His lack of experience in bridge building counted against him, as did his revolutionary design for the span.

Thwarted by practicalities, he turned his attention to perfecting his design.

Claiming to draw his inspiration from a spider's web, Paine sought to liberate bridge design from the restrictions imposed by a semicircle. He concentrated his attention on what geometricians call the "chord of a circle", which, simply put, is a straight line between two points on a circle. Using a chord meant that the height of the arch could be adjusted to the demands of the topology of the area to be spanned.

Goodbye, hump-backed bridges.

Convinced that he had cracked the problem, Paine applied for a patent on his idea, the application being granted on August 26, 1788 (patent No. 1667), specifically for a bridge, using his design, to span the river Don in Sheffield.

Despite having the patent to hand, the project was stillborn.

Desperate to raise some public interest in his design, Paine turned his attention to creating a 110 foot-long iron bridge, effectively a bridge to nowhere, on the bowling green of a public house, the London Stingo, in Lisson Green, on the edge of London's Paddington.

Quite what the bowling fraternity thought of his erection is unrecorded.

Paine had interested Thomas Jefferson in the project. The Sage of Monticello was enthusiastic, convinced that Paine would build an arch of up to five hundred feet and that any bridge so constructed would soon cover its building costs in toll fees generated.

Work was started in May 1790 and completed in the September, eliciting a congratulatory note from Jefferson, "I congratulate you sincerely on the success of your bridge. I was sure of it before from theory: yet one likes to be assured from practice also."

But fine words butter no parsnips.

No money was forthcoming to enable Paine to build a bridge to his new design across a river and, by October 1791, the structure was rusting. Disheartened, Paine suffered the ignominy of seeing his bridge dismantled and packed off to Yorkshire, some of the iron subsequently being used to build a bridge spanning the River Wear in Sunderland in 1796, at 240 feet then the longest iron bridge in the world.

At least, the bowlers of the London Stingo had got their green back.

By then, Paine had weightier matters on his mind. The Pitt administration, fearing a revolution at home, started to crackdown on agitators and dangerous sorts. With a warrant out for his arrest, Paine skipped across the Channel to France in September 1792.

It is a pity that there was no bridge to facilitate his escape.

7. Charles Lindbergh (1902 – 1974)

By any stretch of the imagination Charles Lindbergh was a complex character.

He is best known for his solo, non–stop, trans–Atlantic flight in 1927 from Long Island to Paris in a single-engine plane called *Spirit of America*. Tragedy befell him in 1932 when his son, Charles Junior, was kidnapped and subsequently murdered in what was described by H. L. Mencken as "the biggest story since the Resurrection".

Returning from self-imposed exile in Europe to the States in 1939 at the outset of the Second World War and right up until the Pearl Harbour attack, Lindbergh adopted a prominent anti-interventionist stance, attracting a public rebuke from President Roosevelt and allegations of fascist sympathies.

Once he engaged with the war effort, though, he put his undoubted aviatic acumen to good use, flying over fifty combat missions during the war against Japan in the Pacific region. For the rest of his life, however, he was dogged by allegations of being an eugenist and a philanderer.

The reason for Lindbergh gracing these pages is for his now little-remembered involvement in the development of heart surgery, in particular, the perfusion pump.

Our story begins in 1930, when Lindbergh's sister-in-law developed a heart condition which proved to be fatal. It set him wondering why it was not possible to repair surgically defects in the body's major organ.

He was introduced to the Nobel Prize winning surgeon, Alexis Carrell, who was working on methods to keep organs alive outside of the body. In fact, Carrell had developed a nutrient-rich fluid that did the trick but lacked the technological know-how to ensure that the organ was continuously exposed to oxygenated blood, a process known as perfusion.

This is where Lindbergh came in.

By May 1931, he had advanced sufficiently in his researches to publish, in one of the shortest ever articles to appear in the journal, *Science*, details of a device which circulated fluid constantly through a closed system.

It generated little attention.

By 1935, Lindbergh had come up with a solution to Carrell's problem, a Pyrex pump, consisting of three chambers or, to use his own words, "an apparatus, which maintains, under controllable conditions, a pulsating circulation of sterile fluid through organs for a length of time limited only by the changes in the organs and in the perfusion fluid".

The use of Pyrex was critical.

Pyrex was the only material Lindbergh had found that didn't cause the sort of blood clots and other complications that ordinary glass did. The heart was placed in a slanting tube and the carotid artery was connected to a second, small glass tube. Air pressure would force Carrell's nutritious fluid from a lower chamber through the tube and artery to the heart, gravity then taking over and forcing it back down to the lower chamber. There were no moving parts.

There was one problem; the absence of a filter, an ersatz kidney, meant that the organ's secretion mixed with the fluid from the perfusion pump, requiring it to be changed frequently.

Nonetheless, the duo carried out a public demonstration of their pump on April 5, 1935, perfusing a cat's thyroid gland which, after eighteen days, was still alive and, more importantly, healthy and replicating.

The public response to this breakthrough was phenomenal.

It was described as "the fountain of old age" and some speculated that Lindbergh's contribution would earn him more fame than his aeronautical achievements. The duo even graced the cover of *Time* magazine in July 1935, but the press hysteria proved too much for Lindbergh, forcing him to flee to Europe.

Over the next four years, nearly a thousand trials of the pump were carried out and it never malfunctioned, although the absence of a filter continued to pose the threat of contamination. It was a star exhibit of the World Fair in New York in 1939.

But only around twenty of the pumps were ever produced.

What went wrong?

It was tricky to use and attaching the artery to the glass tube was difficult. It was too easy to tear or damage the artery, making the organ to which it was attached unusable. By 1940 the pump was abandoned.

Despite that, it became the forerunner of surgical devices such as the heart-lung machine and gave surgeons a methodology to work on to stop the heart during an operation.

Lindbergh, though, despite contemporaneous predictions to the contrary, is still better known these days for his flying exploits.

GIVING IT ALL WAY

W HAT IS THE use of living, if it be not to strive for noble causes and to make this muddled world a better place for those who will live in it after we are gone? – Winston Churchill, speech delivered in Dundee, October 10, 1908.

Not every inventor seeks to turn the product of their little grey cells into a money-making machine. Some are imbued with the kind of philanthropic spirit that Churchill so grandiloquently described. Others see their invention as a sprat to catch a mackerel, their seeming largesse designed to encourage the sale of other, more lucrative, products. Some do not recognise the significance of what they have done, while others are such pioneers that there is no infrastructure to protect their rights.

The net result is the same; they miss out on the financial rewards that their inventions merit.

Here are the stories of six individuals who, for a variety of reasons, ended up giving the rights to their inventions away and received little more than a pittance for their troubles.

8. Wilhelm Rontgen (1845 – 1923)

Without question, one breakthrough that has revolutionised the medical profession is the discovery of the x-ray machine.

The process of x-raying a patient in a hospital or in the dentist's chair to get a picture of what is going on inside is so routine that we barely give it a second thought. But someone somewhere must have identified these radioactive rays which illuminate our insides, and this is where the German physicist, Wilhelm Rontgen, comes in.

In 1895, he began investigating what happened when an electrical discharge was passed through various types of vacuum tubes. In early November, he was concentrating on testing a tube created by Philipp von Lenard. Modified by adding a thin aluminium window to allow the cathode rays to escape, it had a cardboard covering to stop the aluminium from being damaged. Despite the cardboard cover stopping light escaping, Wilhelm noted that when a small cardboard screen coated with barium platinocyanide was placed close to the aluminium window, a fluorescent glow could be detected.

On November 8, Rontgen then turned his attention to the Hittorf-Crookes tube, which had a thicker glass wall. Again, he covered it with cardboard and attached it to a coil to generate an electrostatic charge. He darkened the room and as he passed the coil charge through the tube, he noticed a shimmering effect which came from a barium platinocyanide screen he had placed nearby.

It dawned on him that he may have isolated a new form of ray, which he dubbed x, the algebraic notation used to denote an unknown quantity.

Locking himself away in his laboratory for a couple of weeks to examine the properties of these rays, Wilhelm summoned his wife, Anna Bertha, and took the world's first x-ray. When she saw the bones of her hand, his wife exclaimed, "I have seen my death".

It's hard being an inventor's wife.

Experimenting to see which materials had the ability to stop the rays, Rontgen positioned a small piece of lead, while the discharge was occurring. On the screen he saw his own skeleton, flickering and ghostly, the first radiographic image.

He published the first of three papers he wrote on his discovery – *Uber eine neue Art von Strahlen (On a New Kind of Rays)* – on December 28, 1895 and was awarded an honorary doctorate by the University of Wurzburg. In 1901, Wilhelm was awarded the very first Nobel Prize in Physics.

However, financial success didn't follow his landmark discovery. First of all, and nobly, the Nobel prize winner refused to patent his discovery, wanting mankind to benefit from the practical applications of his discovery. He also donated the prize money from the award to his university at Wurzburg.

Philanthropy, however, doesn't make you immune from harsh economic reality. Despite inheriting a fortune on his father's death, two million Reichsmarks, the rampant inflation of the Weimar republic ate into it and shortly after the end of the First World War, he was declared bankrupt.

He saw the rest of his life out quietly at his country home in Weilheim, near Munich, and on his death, in accordance with his wishes, his personal and scientific correspondence was destroyed.

We owe Wilhelm a huge debt of gratitude for allowing humanity to benefit from the discovery of X-rays and not seeking to profit from it.

9. Jerry Siegel (1914 – 1996) and Joe Shuster (1914 – 1992)

"Look! Up in the sky! It's a bird. It's a plane. It's Superman" – lines spoken during the introduction to the American TV series, *Adventures of Superman*, which ran from 1952 to 1958.

Superman is one of the most enduring superheroes. The man from the planet Krypton wandered around Earth as Clark Kent on the look-out for possible trouble and adventure. The stories of his astonishing derring-do in the fight against evil have enchanted millions and filled the coffers of publishers and movie companies over decades.

Yet, even the most ardent comic book fan would be hard pushed to name the man who had invented superman.

So, who created him, and did they get a sizeable share of the income their creation earned?

From around 1933 Messrs Siegel, a writer, and Shuster, an artist, had been developing the idea of a character who would ultimately turn out to be Superman.

Siegel's father, Mitchell, had died on June 2, 1932 during a robbery, staged at his second-hand clothes store in Cleveland. Although the coroner claimed that Siegel's dad had died of a heart attack, the police report indicated that gunshots were heard. In memory of his father, Siegel created a character who was immune to bullets and would wage war against evil. Shuster was corralled to provide the artwork.

Coming up with an idea and selling it are two different things.

The duo's original idea was to sell it to a newspaper syndicate to be run as a cartoon strip with them retaining ownership and rights to the Kryptonite.

Unfortunately, they were unable to find any takers.

By now, they were working for National Allied Publications and struck a deal in March 1938 with their successor company, Detective Comics, Inc. Under the terms of this contract, the duo assigned all rights, goodwill and title to Superman to Detective Comics for the princely sum of $130.

Superman first appeared on the cover of *Action Comics* on April 18, 1938 and was an overnight sensation.

Siegel travelled to New York to meet the co-owner of Detective Comics, Harry Donenfeld, in an attempt to renegotiate the ill-advised contract but was told to go away. The owners did throw the duo some scraps, offering them first refusal on any other creations they may come up with, allowing them six weeks to make a decision.

Siegel presented them with the idea of Superboy, stories of Superman's childhood, but encountered radio silence. True to form, Superboy appeared in *More Fun Comics*, whilst Siegel was serving in the army, the script owing much to Siegel's and the graphics largely based on Shuster's illustrations.

In 1948, the pair launched legal action to regain control of the characters and to get a "just share" of all the profits that had been made out of Superman. They failed to wrest control of the character but did get occasional slices of the action. It was not enough to transform their lives of penury.

Their partial breakthrough came when DC Comics sold the film rights to Warner Brothers in 1975. Anxious to avoid any bad press which might have marred the launch of their block-buster movie in 1978, starring Christopher Reeve, Marlon Brando et al., the film company agreed to pay the pair $20,000 a year and include their names on all credits on future Superman publications.

The story didn't end there.

In 1999, after Siegel had died, his family finally won rights to half of his creation, but that decision was immediately challenged and the only ones who have subsequently got rich out of the whole mess are the lawyers.

Siegel and Shuster gave the world Superman and gave him away for $130.

10. Ruth Wakefield (1903 – 1977)

Now, here's an intriguing question; What would you do to secure a lifetime's supply of chocolate?

Ruth Wakefield ran a restaurant-cum-travel-lodge called the Toll House Inn, situated halfway between Boston and New Bedford in Massachusetts. One of the specialities of the house was a thin butterscotch nut cookie, served with ice cream, known as the Butter Drop Do Pecan Icebox Cookie, a bit of a mouthful but it sounds delicious.

While experimenting with variations on this popular item, Ruth decided to add some chocolate fragments to the batter.

She wanted to use an unsweetened chocolate without milk or flavouring called Baker's chocolate. Finding that she didn't have any, Ruth used chocolate from Nestlé's proprietary semi-sweet bar.

Chopping the bar into small bits, allegedly with an ice-pick, and putting them into the brown sugar dough with nuts, she was surprised to find that the chocolate hadn't melted. Instead, they remained as chunky as when she added them to the mix.

What the enterprising Ruth had created was what we now know as the chocolate chip cookie, but back then it was known as the Toll House Chocolate Crunch Cookie. In 1938, Ruth added the recipe to her best-selling cookbook, *Toll House Tried and True Recipes*. In truth, the chocolate chips do melt but they retain their shape because of the way the fat in the chocolate is aligned.

In her recipe book Ruth explained her methodology. "At Toll House we chill this dough overnight. When ready for baking, we roll a teaspoon of dough between palms of hands and place balls two inches apart on a greased baking sheet. Then we press balls with finger tips to form flat rounds. This way cookies do not spread as much in the baking and they keep uniformly round."

The cookies sold like hot cakes in the New England area and the recipe was published in a Boston newspaper.

Noticing that the sales of his semi-sweet chocolate bars had rocketed since the development of the cookie, Andrew Nestlé, of the eponymous company, decided to make an unusual agreement with Ruth. He secured the right to print the recipe for the Toll House Cookie on the package of his chocolate bars, reportedly paying the princely sum of one dollar for the privilege. To sweeten the deal still further, Ruth was offered a consultancy position to assist in the development of further recipes.

Her fee for this?

Free chocolate for the rest of her life. Sounds like a good deal.

In what we would now call a series of re-branding exercises, the Toll House Cookies became Nestlé Toll House Real Semi-Sweet Chocolate Morsels and then chocolate chips. Their popularity extended beyond New England, principally because during the Second World War soldiers stationed in the area spread the word and took the delicious biscuits home with them.

It soon established itself as America's favourite cookie and Nestlé, thanks to the unusual agreement struck with

Ruth, owned the rights and, more importantly, took all the profits from the cookie.

This cosy arrangement was interrupted in 1983, when a federal judge ruled that there were ambiguities in the deal struck in 1939 and that Nestlé was no longer entitled to retain exclusive rights to the Toll House trademark.

So, the market was opened up.

It is estimated that around seven billion chocolate chip cookies are eaten in America a year, 50% of which are home-made.

Ruth sold the Toll House inn in 1967, it burned down on New Year's Eve in 1984, and retired to Duxbury in Massachusetts, where she died in 1977.

She may have invented America's favourite cookie and given it away for a dollar, but at least Ruth got thirty-eight years' worth of free chocolate!

11. David M Smith (c1940s to present)

Ah, the nineteen eighties. Whether you loved them or hated them, the music of the time was dominated by synthesisers and electronica. One of the developments that made this possible was the creation of the MIDI or Musical Instrument Digital Interface in 1983.

It quickly became the universal standard and you would have thought that its inventor would have found the key to a fortune, but you would be wrong, as the story of Dave Smith reveals.

The problem with electronic instruments pre-MIDI was that they couldn't talk to each other. OK, a clever keyboard player could play two instruments at the same time, one with their left hand and the other with their right, but that was pretty much as far as it went.

A graduate in Computer Science and Electronic Engineering from UC Berkley, Smith was fascinated by synthesisers, setting up his own firm, Sequential Circuits, in 1974, and developing, in 1977, the Prophet 5, one of the first analogue polyphonic synthesisers.

Sequential Circuits went on to become one of the most successful synthesiser manufacturers ever.

But Smith wasn't satisfied.

He wanted to develop a protocol whereby electronic instruments and synthesisers could communicate, allowing the musician to control a range of instruments from one synthesiser or computer.

In 1981, he issued a challenge to the industry to back a universal protocol.

To set the ball rolling, he created a rough draft of what it might look like, calling it the Universal Synthesiser Interface. Few came forward to enlist, but one who did was Ikutaro Kakehashi, the founder of Roland. The two collaborated during 1982, communicating, it seems astonishing to write this these days, by fax. Smith was ready to reveal what they had come up with, by the time of the National Association of Music Merchants conference in 1983.

By today's standards, the specification for MIDI was fairly rudimentary, consisting of eight sheets of paper and limiting itself to a set of basic instructions you might want to send between two synthesisers, like which notes to play and at what volume.

But it worked. Smith was able to link up his Prophet 600 synthesiser with a Roland JP-6. A musical revolution had begun.

MIDI's development coincided with the development of the PC, whose processors were now fast enough, using MIDI, to sequence notes and to control a number of keyboards and drum machines operating at the same time. It also allowed aspiring musicians to operate at home rather than spending time in expensive recording studios. It didn't stop there. MIDI technology has been incorporated in Mac OS since 1995 and is used in your smartphone, powering the first wave of ring tones. Games like *Guitar Hero* use it.

It has stood the test of time. The basic protocol has been added to but remains the same.

Why did Smith not make a fortune?

Well, he gave it away. Explaining what may seem, on the face of it, a baffling decision, he said, "we wanted to be sure we had 100% participation, so we decided not to charge any other companies that wanted to use it".

Very magnanimous.

On the other hand, it may have been a sprat to catch a mackerel, making products such as synthesisers more valuable and desirable. But even then, Smith had to sell Sequential Circuits to Yamaha in 1988 to stave off bankruptcy.

Smith is still making and selling synths with his own company, Dave Smith Instruments, but by refusing to license MIDI in order to make it freely available, he missed out on a fortune.

12. Daisuke Inoue (1940 – present)

Sometimes there are people who are so unworldly that they do not recognise the importance or potential of what they have done. Daisuke Inoue is one of those people.

Personally, I hate karaoke and will run a country mile from any establishment that is advertising a karaoke session. The prospect of being in the same enclosed space as some inebriated people, who cannot miss the opportunity to belt out, out of key and out of time, some hit song fills me with dread. I am put in mind of Samuel Taylor Coleridge's epigram, *On a Volunteer Singer;* "swans sing before they die; 'twere no bad thing/ did certain persons die before they sing". But there is no denying that it is a popular phenomenon and it is all down to Daisuke.

Born in Osaka on May 10, 1940 Daisuke started playing the drums at school because it seemed the easiest instrument to play. Cue old joke; What do you call someone who hangs around musicians? A drummer.

He wasn't very proficient but got gigs in bars in Kobe, accompanying businessmen who wanted to sing to a crowd. Since he couldn't read music, Inoue had to follow the singers and soon his off-beat drumming style proved popular, with them at least.

In 1971, he was asked to accompany a businessman on tour. Unfortunately, Daisuke was unable to go and so began to wonder what he could do to provide his client with the musical backing he needed.

The answer was simplicity itself but, hey, that is what a moment of genius is. If Daisuke couldn't be there in person, banging the drums, then why could he not be there in spirit? In other words, he could lay down the rhythm and basic accompaniment on a tape, which the businessman could play while he was warbling away.

The idea of karaoke was born.

Inoue developed eleven karaoke boxes, karaoke means empty orchestra, which contained 8 backing tracks for would-be singers. He rented them out to bars in Kobe and they proved to be phenomenally successful.

However, exhibiting the characteristically flawed genius of many in this book, Daisuke made one catastrophic mistake. He forgot to patent his invention, thinking that he hadn't really invented anything. All he had done, he modestly thought, was put together something from items of equipment that already existed, missing the point that what he had created was novel and had never been done before.

Daisuke and his machines had caused a stir and karaoke sessions were becoming increasingly popular. The buzz that had been created came to the attention of some of the large Japanese corporations. They probably couldn't believe their collective luck when they discovered that Daisuke had not patented his machine.

Never ones to look a gift horse in the eye, the big corporations seized their opportunity. It is estimated that Daisuke's failure to patent the karaoke machine meant that he lost out on around $110 million in royalties.

Some oversight.

Daisuke did, however, profit marginally from the karaoke boom that he created. He invented a pesticide that repelled cockroaches, rats, and other vermin that, showing good taste, in my view, attacked the electrical wiring in the machines.

He also got some gratuitous recognition by being named one of the most influential Asians of the century by *Time Magazine* in 1999 and having a Japanese film called *Karaoke* made about him in 2005.

Clearly, the moral of his story is never under-estimate what you have done.

13. Stephen Foster (1826 – 1864)

At primary school, for some unaccountable reason as it was situated in a county with a fine folk tradition, the songs we sang were mainly American.

One particular favourite, which we sang with gusto, was *Camptown Races* which started off, "Camptown ladies, sing this song/ doo-da doo-da/ The Camptown racetrack's five miles long/ Oh doo-da day." It sounded better than it appears on paper. It was one of over two hundred songs written by Stephen Foster, not that I knew at the time nor, frankly, cared.

Foster has been called the father of American music and many of his songs are popular to this day. In his musical canon are ditties such as *Old Folks at Home*, *My Old Kentucky Home*, *Jeanie with the light brown hair*, *Old Black Joe* and *Beautiful Dreamer*.

As a youngster, Foster joined a quasi-secret society known as the Knights of the Square Table, who spent their evenings singing songs. He was heavily influenced by a German musician, Henry Kleber, who ran a music store in

Pittsburgh, and Dan Rice, an itinerant entertainer. It was during this period that he wrote one of his most famous songs, *Oh! Susanna*, although the first song he published, at the age of eighteen, was *Open Thy Lattice, Love*.

When he was twenty-four and married, Foster decided to earn his living as a professional song-writer.

The problem with being the first in your field is that there are usually no rules of engagement. So, whilst Foster would generally find someone who would pay him some money for the rights to publish his songs, there was no such thing as an established music royalty system.

Oh! Susanna, published in 1848, and the unofficial anthem of the Californian gold rush, earned him just $100 while his publisher raked in $10,000.

Returning to Pennsylvania in 1849, he signed a contract with the Christy Minstrels and over the next five years or so wrote many of his most well-known songs, including *Camptown Races* in 1850. They were often in the blackface minstrel style, which was popular at the time. but with subtle changes, as Foster wrote, "to build up taste... among refined people by making words suitable to their taste, instead of trashy and really offensive words which belong to some songs of that order".

But instead of the millions that his works would have earned him these days, he received little more than $15,000 in total for all the songs, which are now the staple of the American songbook.

Inevitably, Foster hit hard times.

His annus horribilis was almost certainly 1855, when he separated from his wife and both his parents died. He reacted to his troubles in the only way he knew how, by writing another hit, *Hard Times Come Again No More*.

What certainly did come no more was money and he was reduced to living a rootless existence, dossing in hotels in New York.

In January 1864, Foster contracted a fever and was severely weakened. His then writing partner, George Cooper, found him, naked, lying in a pool of blood, having hit his head on a wash basin. He died in Bellevue Hospital three days later, on February 13, aged just thirty-seven. In his wallet was found a scrap of paper with the words, "Dear friends and gentle hearts" and just thirty-eight cents. *Beautiful Dreamer*, perhaps his most famous song, was published posthumously.

Part Two

Discrimination

SEXUAL DISCRIMINATION

SEEK TO BE good, but aim not to be great;/ A woman's noblest station is retreat – Lord George Lyttelton, *Advice to a Lady*, 1731.

Very few of us, in what we like to delude ourselves in to thinking are enlightened times, would be bold enough to spout the sort of sexist claptrap that the British politician, Lord George Lyttelton, came up with in his poetic advice to Belinda. Regrettably, though, they were representative of the patronising attitudes men have adopted towards women down the centuries. Fortunately, the tide is turning and women are, at long last, beginning to take their rightful place in society and competing on equal terms with men.

But it has been a long struggle.

Take education. In the eighteenth century women weren't supposed to worry their pretty little heads with the likes of Latin and Greek, the stuff of a gentleman's education. What was good enough for them was mastery of such things as embroidery and knitting. However, not all women were consigned to a life of ignorance. Anna Barbauld, who went on to become a poet and literary critic at a time when it was rare for women to become professional writers, was persuasive enough to get her father to teach her some Latin and Greek. Even then, though, she thought that a formalised education system was unnecessary for the fairer sex. Instead, "the best way for a woman to acquire knowledge", she opined, "is from conversation with a father, or brother, or friend", male, of course.

This rather curious point of view lead to the creation of the Blue Stocking Club, where women could converse, on equal terms, with the great and the good of the male gender. It was parodied mercilessly by the cynics amongst Georgian society but, perhaps, it sowed a seed. After all, the Reverend Sydney Smith paused, in 1810, to consider "why a woman of forty should be more ignorant than a boy of twelve."

But it was not until the twentieth century, and some way into it, that women were allowed to take degrees.

The inventive streak, though, is no respecter of gender, race or creed. Here are the tales of six women who had to battle against sexism, institutionalised or personal, to get their inventions or discoveries taken seriously.

They were not always successful.

14. Lise Meitner (1878 – 1968)

Being Jewish, a woman in academia, and living in Austria in the 1930s weren't the best cards to be dealt with in life and so it proved for nuclear scientist, Lise Meitner.

Born in Vienna, Lise was only the second woman to be awarded a degree in Austria. To further her studies, she moved to Berlin where she met Otto Hahn and found a position, a cupboard next to a lab and working as a guest without remuneration, at the Kaiser Wilhelm Institute for Chemistry.

It was only when she was offered a paid position elsewhere that her role at the Institute was regularised. Despite these handicaps, by 1917 she and Hahn had discovered a new element, protactinium.

In the aftermath of the First World War scientists around the world were racing to give their respective countries an edge by discovering an element heavier than uranium. Naturally, it was to this problem that Meitner and Hahn next applied their not inconsiderable grey cells. They noticed that whenever they put a neutron on to a heavy Uranium neutron, as you do, they ended up with something lighter. Whilst Hahn carried out the experiments, it was Lise who came up with the explanation for this phenomenon and realised the import of what they had discovered.

The answer was what we now call nuclear fission.

What was happening was that the neutron was splitting into two parts, unleashing a phenomenal amount of energy in the process. It was this energy, which was later harnessed to produce nuclear bombs.

By this time, 1938, the Germans had taken over Austria, in what was known as the Anschluss, and, sensibly, Lise had made good her escape to Sweden. Now that he had the rational explanation to the phenomenon that they had observed, Hahn wrote up the findings and published a paper, ignoring the contributions that Lise had made. In fact, he omitted her altogether.

Some kindly souls argue that the omission was due to political pressure exerted because of the race and gender of Hahn's accomplice. Whether this was the case or whether Hahn just grabbed the glory for himself, we will never know. To add salt to the wound, in 1944 the Nobel Prize in Chemistry was awarded to Hahn alone for the discovery of nuclear fission.

Not unsurprisingly, Lise was royally pissed off.

She wrote, "I have no self confidence... Hahn has just published absolutely wonderful things based on our work together ... much as these results make me happy for Hahn, both personally and scientifically, many people here must think I contributed absolutely nothing to it and now I am so discouraged."

Worse still, she was horrified to find that the first use of nuclear fission was to make an atomic bomb and was devastated when the *Enola Gay* dropped its load on to Hiroshima.

To complete her air-brushing from history, the apparatus that was used to carry out the experiments that led to the discovery of nuclear fission was displayed in Germany's leading science museum for thirty-five years without mentioning Lise's name and or her role in the experiment.

Lise continued with her researches after the war and helped produce one of the first peacetime nuclear reactors. During the course of her career she published some 128 articles.

It was only in the mid–1960s that the enormity of her contribution to the discovery of nuclear fission was recognised. Posthumously, in 1992, she had an extremely radioactive synthetic element named after her, Meitnerium (atomic number 109).

At least the Periodic Table bears testament to her brilliance.

15. Cecilia Payne (1900 – 1979)

The stars I see twinkling at night on the few occasions they are not hidden by clouds are a constant source of wonderment to me. Those of a more enquiring mind might wonder what they are made of and a few, a very few, would take the trouble to find out. One such was the British-born astronomer and astrophysicist, Cecilia Payne.

Her contribution to our understanding of stars should have assured her a stellar career but it was for decades hidden under the penumbra of male chauvinism that pertained in the groves of academe at the time.

Cecilia was a bit of a brain-box and read botany, physics and chemistry at Newnham College in Cambridge in the early 1920s but she did not get a degree as the university only started awarding them to the fairer sex in 1948. She did, however, listen to a lecture by Arthur Eddington, which sparked her nascent interest in astronomy.

Winning a scholarship, Cecilia moved to the United States in 1923 and enrolled in the graduate programme run by Harvard College Observatory, specifically established to persuade women to study there. She was encouraged to write a doctoral dissertation and in 1925, Cecilia became the first woman to receive a PhD from Radcliffe College, now part of Harvard. Her dissertation was entitled *A Contribution to the Observational Study of High Temperature in the Reversing Layers of Stars.*

And some contribution, it was too.

I will not bore you with the details, the precise findings and analytical processes that she used go way over my head, but, in essence, Cecilia concluded that whilst the stars shared the same elements to be found in the Earth, hydrogen, in particular, and helium, to a degree, were the most abundant elements in the composition of the stars and that hydrogen was the most predominant element in the Universe. Later, astronomers were to call her work "undoubtedly the most brilliant PhD thesis ever written in astronomy".

Cecilia's problem, however, was that she had made her discovery in 1925 and it flew against the then received wisdom that the composition of sun and the stars was no different from that of the Earth.

The villain of the piece, Henry Norris Russell of Princeton University, now enters our story. He was assigned the task of reviewing Cecilia's dissertation. Because the findings were contrary to the commonly accepted theories, he declared them "clearly impossible" and Cecilia, bowing to the pressure exerted by the eminent professor, amended her conclusions, stating that the predominance of hydrogen and helium, which she had calculated, was "almost certainly not real".

But something about Payne's conclusions intrigued Russell and he conducted his own investigations, concurring four years later in 1929, in a short paper, with the proposition that the principal constituent of the sun and stars was hydrogen. Russell magnanimously acknowledged Payne's contribution but, in popular and academic circles, he was recognised as the person that had established this ground-breaking fact.

Cecilia spent most of her career studying stars but was forced by the conventions of the time to accept low paid, low grade academic positions. It was only in 1956, that she was able to break through the glass ceiling, when she was appointed a professor at Harvard.

To add to the irony, Cecilia was awarded the Henry Norris Russell Prize for her contributions to astronomy in 1976. She was typically phlegmatic, commenting at the time, "the reward of the young scientist is the emotional thrill of being the first person in the history of the world to see something or to understand something".

Quite.

16. Nettie Stevens (1861 – 1912)

How the sex of a child is determined at conception has puzzled many grey cells more powerful than mine over the centuries.

Aristotle thought that it was all about environmental heat and advised males who were looking to sire sons to copulate in the summer. A popular theory going the rounds in Europe during the nineteenth century was that it was all about nutrition. A good diet produced girls whilst a poor one resulted in males. That was one way of keeping down the food bill.

A more drastic course of action was promulgated by the eighteenth century French anatomist, Michel Procope-Couteau (1684 – 1753), who, in *The Art of Having Boys,* revived Parmenides and Anaxagoras' theory that the testicles and ovaries were either male or female. Excision of the unwanted reproductive organ would ensure the birth of a child of the desired sex. I'm not sure too many followed his strictures but he did later come up with a more practical alternative.

The female should lie on the correct side and let gravity take care of the rest.

It was only at the turn of the twentieth century that we had a clearer idea of how sex was determined and this is where some insects and Vermont-born geneticist, Nettie Stevens, come in.

A late entrant into the groves of academe, she was awarded a doctorate in cytology by Bryn Mawr College in Pennsylvania in 1900 and continued as a researcher, looking into the subject of sex determination.

Drosophila melanogaster, to give the fruit fly its Latin tag, is often used in research because they can be bred readily and quickly in laboratory conditions, lay a lot of eggs and have a simple genetic make-up. Of particular interest to our Nettie was the fact that they only have four sets of chromosomes and it was these that she studied under her microscope in 1905.

She quickly discovered that the chromosomes differed between the sexes.

Transferring her attentions to the mealworm, Stevens identified and isolated a chromosome she called Y, realising that it was linked to and the opposite of the X chromosome, discovered and so named by Hermann Henking in 1890.

Extending her research to include egg tissue and the fertilisation process, Nettie realised that the X and Y chromosomes always existed in pairs and that it was the presence or absence of the Y that determined the gender of the result of the fertilisation process. The sex of a baby had nothing to do with environmental factors, it was down purely to genetics and the Y chromosome.

Nettie was not working in a vacuum.

Edmund Wilson was also carrying out researches into how sex was determined. His methods differed from Nettie's, concentrating on species where the male had one fewer chromosome than the female and on the testes, as eggs were too fatty for his staining methodology.

It is almost certain that Wilson had access to Nettie's results and although he concluded that environmental factors also had a hand in sex selection and was less adamant in its conclusions, his paper was published first and being a man, he was credited with discovering the chromosomal basis for sex determination.

The other villain of the piece is the prominent geneticist, Thomas Hunt Morgan.

He wrote the first text book on genetics and there is evidence that he corresponded with Nettie, asking for more and more details of her experiments. When she died of cancer in 1912 Morgan was dismissive of her contribution, inferring she was more of a researcher than a scientist. There was no mention of Stevens in his magnum opus and to make matters worse, in 1933 Morgan and Wilson were awarded the Nobel Prize in Physiology or Medicine for "their discoveries concerning the role played by the chromosome in heredity".

Although Stevens' theory could not be proven at the time, it turned out to be right and it is only now that her contribution is beginning to be recognised. It is truly ironic that sex discrimination should blight the story of the understanding of how sex is determined at conception.

17. Rosalind Franklin (1920 – 1958)

I am a great fan of crime stories from the period between the two World Wars, known as the golden age of detective fiction. Policemen and amateur sleuths had to rely on their wits and their powers of analysis, reason and deduction to solve many a fiendish crime which, at first blush, seemed both impossible to have been committed and to crack. They invariably did, though, usually because the felon left some tell-tale sign that led to their undoing.

Life has moved on and these days the police have a more powerful array of tools at their disposal, at least if you believe the police dramas which are the staple fare of our TV screens, not least DNA testing. If I had even the faintest inkling to commit a crime, the threat of being unmasked by my DNA would be enough to put me off. Interestingly, the tale of the discovery of DNA is a murky one with elements that would not have been out of place in a good whodunnit.

Rosalind Franklin always wanted to be a scientist, even though her father tried to steer her away from a career path that was nigh-on impossible for women to make much progress in. She was fortunate enough to attend St Paul's Girls' School, one of the few schools at the time that taught physics and chemistry to girls, and then graduated from Newnham College, Cambridge.

During the Second World War, Rosalind studied the structure and uses of coal and graphite, publishing several papers and contributing to the development of more effective gas masks. She was awarded a PhD in Physical Chemistry by Cambridge University in 1945.

After the war, Rosalind went to Paris to work under Jacques Mering, from whom she learned about the use of x-ray diffraction techniques to explore and understand the molecular and atomic structures of crystals. Then, in 1951, she made the fateful decision to accept a three-year research scholarship at King's College, London.

Maurice Wilkins was trying to understand DNA by using X-ray crystallography and so Rosalind was perfectly equipped to contribute to the project. But Wilkins, who was away when Rosalind arrived, assumed that she was hired help rather than be someone who could more than contribute to the project.

Their relationship never recovered from this rocky start.

Working with a student, Raymond Gosling, Rosalind continued to refine her X-ray images of DNA fibres, using ever finer strands. Wilkins, in somewhat of a huff, spent increasingly more time with his friend, Francis Crick, at the Cavendish Laboratory, where Crick and James Watson were attempting to understand the structure of DNA by using a model-based approach.

Around this time Rosalind made a dramatic discovery when looking at what later became known as Photo 51. The DNA in the image had a distinct helical structure with two strands attached at the middle. She gave details of her findings in a lecture but no one seemed to pay any notice.

However, at a conference, at which Crick and Watson rolled out their theories about the structure of DNA, Rosalind challenged them, pointing out that she was working with empirical data not highfalutin ideas. This open criticism of his friends worsened relationships between Wilkins and the woman he now called the Dark Lady. Sensibly, Rosalind decided to move on and took a position at Birkbeck College in 1953.

But during the move, Wilkins came into possession of the famous Photo 51, certainly without Rosalind's permission, and showed it to Crick and Watson. It was an earth-shattering moment. Here was the missing piece of information, which Crick and Watson needed to complete their accurate model and proof positive that DNA's helical structure had two strands attached in the middle by phosphate bases.

The duo rushed to print, publishing an article on the structure of DNA in a 1953 edition of the scientific journal, *Nature*. Ironically, the same edition carried articles by Wilkins and Franklin on the X-ray data they had compiled about DNA but it gave the impression that their contribution was supplementary to rather than one that had informed Crick and Watson's discovery.

Rosalind continued her researches at Birkbeck, now turning her attention to the structure of tobacco mosaic virus before succumbing to cancer, which she may well have contracted through her work with X-rays.

In 1962, Crick, Watson, and Wilkins were awarded the Nobel Prize in Physiology or Medicine. There was no mention of Rosalind and it is only recently that her contribution to the understanding of DNA has been acknowledged. The Nobel Prize, of course, cannot be awarded posthumously.

Did Crick steal the photograph? Perhaps we should run a DNA test.

18. Margaret Knight (1838 – 1914)

Born in York, Maine, Margaret worked in a cotton mill as a child. At the tender age of twelve she witnessed an accident in the factory, where a steel-tipped shuttle shot out of a mechanical loom, stabbing a work colleague. Within weeks of witnessing this traumatic event, she had invented a safety device which prevented a recurrence. Margaret never patented her invention, which was soon adopted by other mills. Indeed, quite what it was is not clear, it might have been a guard to stop the shuttle from flying off or some kind of safety device to stop the loom. Either way, Margaret had made a valuable contribution to safety in the mills but never saw a penny for her initiative.

Dogged by poor health, Margaret left the mill before she reached the age of twenty.

In 1867, she moved to Springfield, Massachusetts and started working at the Columbia Paper Bag Company.

The industrialisation of paper bag manufacturing had taken a major leap forward when, in 1852, Francis Wolle, a Moravian priest-cum-schoolteacher-cum-business man from Pennsylvania, invented and patented a paper bag-making machine.

The basics of its design are still used today. That said, the bags it produced were fairly rudimentary, rather like envelopes and quite flimsy, without the flat bottoms that are used today for takeaways and the like. How to improve the paper bag-making machine was a challenge which Margaret could not resist.

She spent time working on a device that would cut, fold and paste the bottoms of bags. When her employer complained about the time Margaret was spending on developing her prototype, she offered him the rights, for a price, if she could come up with a solution.

He agreed and after knocking out thousands of bags on her wooden model, Margaret was satisfied that she had a fully functional device. She had a metal prototype made in Boston, a requisite for submitting a patent application.

This is where her problems began.

A man called Charles Annan had recently visited the factory and had paid particular attention to her prototype. So meticulous were his observations that he filed for a patent for a machine, which looked suspiciously like the square bottom paper bag-making machine that Margaret had painstakingly developed and trialled.

Our heroine wasn't going to let this device slip from her grasp and filed a patent interference suit against Annan. With the bit firmly between her teeth, Margaret spent upwards of $100 a day plus expenses in garnering depositions from herself and other key witnesses in preparation for the trial.

As part of his defence, Annan claimed that because Margaret was a woman, she could not possibly understand the complexities of a machine like this.

Margaret's preparation paid off, though, her notes, diary entries, samples, and affidavits convincing the court to dismiss Annan's rather chauvinist arguments and to find in her favour. However, it had taken three years to get that far.

Establishing the Eastern Paper Bag Company, she began to receive royalties for her invention.

Margaret then became a bit of a serial inventor, credited with around ninety inventions and holding twenty-two patents. Her inventions included a new window frame and sash design, a numbering machine, an automatic boring machine, and a spinning or sewing machine.

Although these all made her money, by the time she died in 1914, she had just $300 to her name.

19. Martha Coston (1826 – 1904)

For a century or more the Coston flare system was the usual way by which ships could communicate with the shoreline and vice versa. Indeed, its use was a requirement of marine insurers.

The signals were produced in the form of cartridges, which were fired into the air from a signal-pistol. There were three colours, white, red, and blue, and by sequencing them, a rudimentary form of messaging, akin to semaphore but one that could be used at night, was developed. The light emanating from the pistol was so bright that the signaller was advised not to look at it. The point, of course, was that it could be seen from a distance.

So, whose brainwave was it?

Step forward, Martha Coston, whose tale is one of triumph over adversity. If being widowed at the age of 21, in 1848, with four children to look after was not enough, further misfortune befell her when two of her children and her mother died shortly afterwards. She needed to find a way of supporting herself. Going through her dead husband's papers, Martha found that he had been working on a system for signalling at night. Benjamin's papers consisted of plans and chemical formulae and, whilst there was a kernel of an idea there, a lot of work would be needed to bring it to reality.

Indeed, it took ten years of hard work for Martha to create a workable system. As she wrote, "The men I employed and dismissed, the experiments I made myself, the frauds that were practiced upon me, almost disheartened me; but … I treasured up each little step that was made in the right direction, the hints of naval officers, and the opinions of the different boards that gave the signals a trial. I had finally succeeded in getting a pure white and a vivid red light."

Needing a third colour, the breakthrough came when she was watching a firework display in New York City to celebrate the completion of the transatlantic telegraph cable in 1858. The blue fireworks were particularly luminous and visible; a blue flare would complete her system.

On April 5, 1859 a patent, number 23,536, was granted for a night signalling system. Sadly for Martha, the inventor on the patent was named as Benjamin, her involvement being relegated to that of administratrix of her husband's estate.

The US Navy were interested in the flare system and placed an order for $6,000 worth of flares from the Coston Manufacturing Company, which Martha had established. She then went on an extended tour of Europe, getting patents for her invention in England, France, Italy, Denmark and the Netherlands. Returning to the US in 1861 at the outbreak of the American Civil War, Martha persuaded Congress to buy the US Patent to her invention but they were only prepared to pay $20,000 rather than the $40,000 she wanted.

The US Navy used the Coston flare system extensively during the conflict and it was particularly instrumental in coordinating efforts in the battle of Fort Fisher in 1865 and in spotting blockade runners. As the Coston Manufacturing Company were selling the flares at less than cost price, after the war Martha calculated that the government owed her $120,000 in compensation. With some reluctance, they offered her a measly $15,000.

In 1871, Martha was awarded a US patent in her own right, number 115,935, for improvements to the night signalling system and, by the mid-1870s, all the US Life Saving Service stations were equipped with Coston flares. Martha also sold her signals to navies, shipping companies and yacht clubs around the world.

Business boomed until the adoption of ship radios.

Martha said she always had to be "ready to fight like a lioness" against chauvinism, prejudice and attempts to rip her off. She persevered and ultimately prevailed.

RACIAL
DISCRIMINATION

I WAS RAISED to believe that excellence is the best deterrent to racism or sexism. And that's how I operate my life – Oprah Winfrey

In her own way, Oprah Winfrey has been extraordinarily successful but, perhaps, her paean to the ability of excellence to overcome all obstacles is a little too optimistic and a tad naïve. It is a depressing testament to the world we live in that the spectre of discrimination on the basis of creed, colour or gender is raising its ugly head again, as it invariably does in times of economic uncertainty.

Abraham Lincoln described slavery as "wrong, morally, socially and politically," but its impact on American society both before, and after, the Civil War, was profound.

Inevitably, the dead hand of institutionalised racism and the consequences of slavery impinged on the world of the inventor. Here are the stories of six individuals who had to battle, some more successfully than others, with these forms of discrimination in order to get their ideas to a wider public.

20. Benjamin T Montgomery (1819 – 1877)

There are many obstacles that beset the inventor but perhaps the one they can do least about is the accident of birth.

The story of Benjamin T Montgomery is, ultimately, one of triumph over adversity but some of the rewards that were due to him for his ingenuity and inventiveness were denied because of his origins.

His problem?

Benjamin was a slave.

Born into slavery in Virginia in 1819, he was sold in auction in 1837 to a plantation owner from Mississippi, Joseph Davis. Unusually for the age, he had learned to read and write, which resulted in Davis giving him enhanced responsibilities, allowing him to operate as a merchant and business manager on the plantation, For all that, though, he was still a slave and this was to become the root of his problems.

Benjamin was an inquisitive individual and had an inventive streak. He is said to have come up with many an invention, although precisely what and how many have been lost in the mists of time.

What is for certain is that in the late 1850s he came up with a neat idea for a propeller for a steamboat, which had to navigate shallow waters. The waters by the plantation, in which he worked, were particularly shallow and it is tempting to speculate that the idea came to him as he watched these large vessels struggle and strain to dock alongside the jetty.

The idea was relatively simple. Instead of having a fixed head, the propeller had blades which could be adjusted and angled to suit the conditions, and, by entering the water at an angle, the blades used the power available to them much more efficiently and provided more manoeuvrability than the standard propellers used at the time. It was not a new concept but was a marked improvement on the propellers patented by John Stevens in 1804 and John Ericsson in 1838.

Encouraged by the fact that it worked, Benjamin applied for a patent in 1858 but, on June 10 of that year, the Patent Office turned it down.

Why?

It was all down to a decision handed down by the Supreme Court in the case of *Dredd Scott v Sanford* in 1857. In a nutshell, the court held that slaves could not have any standing in court for the simple reason that they were not citizens, they were the property of others.

The Head of the Patent Office at the time, Joseph Holt, applied this interpretation to the particulars of the prevailing patent law, which required the putative inventor to swear as a citizen of the United States that the invention had been conceived by them. As Benjamin was not a citizen, Holt concluded, he could not swear the oath.

Joseph Davis and then his brother, Jefferson, later to become President of the Confederacy, applied for the patent on Benjamin's behalf but their applications were turned down because they were not the true inventors of the steam propeller.

On assuming the Presidency of the Confederacy, Jefferson Davis did sign a law, allowing slaves to receive patent protection for their inventions, but that was particular to the Confederacy.

When Benjamin, by now a free man, applied for a patent once more, on June 28, 1864, it was turned down again.

Benjamin did win out in the end. Joseph Davis allowed his slaves to earn money by running businesses, provided that they paid him an amount commensurate with the work they would have done on the plantation. Benjamin's propellers proved successful, enabling him to earn some money from his invention.

In 1862, at the height of the Civil War, the Davis family fled their plantation at Davis Bend, leaving Benjamin in charge.

Despite being under fire from both sides, the plantation survived and Benjamin bought it off Davis in 1866 for $300,000, with the help of the money his propeller had generated and a long-term loan.

In 1867, Benjamin was elected as justice of the peace for Davis Bend, becoming the first African American official in the state of Mississippi. To top that, his cotton was voted best in the world at an International Exposition in 1870.

It had been a long, hard road to get there.

21. Lewis Temple (1800 – 1854)

One of nature's most impressive and deeply memorable sights is a whale coming to the surface of the ocean and evacuating air through its blowhole, before taking another enormous gulp of air and descending once more to the deep.

The majesty and grace of Earth's largest mammal has spawned a new form of tourist industry, whale watching. It is estimated that whale watching generates over $2 billion in tourist revenue per annum and creates employment for some 13,000 people around the world.

Whale hunting, on the other hand, is very much frowned upon these days and rightly so. But it has a long and ignoble history, the earliest recorded hunting scenes dating back to as long ago as around 6000 BCE, according to depictions found in the Neolithic site of Bangudae in South Korea.

The whaling industry was one of America's principal industries in the nineteenth century. In the 1840s, of the 700 or so whaling ships that scoured the oceans for their prey, more than 400 set out from the Massachusetts' port of New Bedford.

Although some ate whale meat and the creatures' bones were a mainstay in the construction of corsets, what was really prized was the oil that could be extracted from its blubber. It was used to make soap and as an illuminator for lamps, cheaper than other products but one which gave off a distinctive and unpleasant smell.

A flavour of the industry at the time and the hold that the creature could have over their pursuers can be gained from the pages of Herman Melville's *Moby Dick*. Successful whaling captains could make a very decent living. To show off their wealth they built big houses in the best neighbourhoods of New Bedford. Not for nothing was New Bedford known as "the city that lit the world".

Killing a whale, though, was very much a hit or miss affair. As well as having to contend with prey that was enormous, strong and reluctant to give up its life easily in dangerous conditions, the whalers were bedevilled by poor technology. Their weapon of choice, a harpoon, once it had entered the flesh of the whale, tended to work loose as the mammal thrashed about. It often meant that even though the whalers had struck their target, the harpoon would not stay in place long enough for them to make the kill.

Lewis Temple was born in Richmond, Virginia, into slavery. In 1829 he moved to New Bedford, where he lived as a free man. He was a skilled blacksmith and established a business selling impedimenta to whalers in Walnut Street on the waterfront of his adopted town. He was also a prominent abolitionist, elected in 1834 as vice president of the town's first anti-slavery society, the New Bedford Union Society.

Many a whaler, who visited Lewis' shop to have their tools and harpoons made or repaired, would bewail the efficiency of the harpoon. This set Lewis thinking and by 1848, he had come up with a revolutionary new design for the harpoon, which was later to be known as Temple's Toggle Iron.

The way Lewis chose to modify the harpoon was to replace the fixed head with one that was moveable.

What this meant was that when the head entered the whale's flesh, it stuck fast and could only be removed by cutting it out. Seasoned whalers were sceptical as to its efficacy but trials soon established the superiority of Lewis' design. It revolutionised the whaling industry, significantly improving the success rate in killing a whale once the harpoon had been driven home.

Lewis never patented his invention, perhaps because he was African American or maybe he didn't get around to it. Although he developed a tidy business from his enterprise, the lack of a patent meant that other harpoon manufacturers were free to use his design.

And they did.

One New Bedford company, Delano and Pierce, did, though, have the decency, in 1853, to fund the construction of new premises for Lewis, at the bottom of School Street, but it was never completed and Lewis didn't get the chance to stand behind the counter of his new establishment.

Disaster had struck.

One evening in the autumn of 1853 Lewis was walking home. Whilst crossing over a plank over an open sewer trench near the construction site of his new shop, he fell off. The injuries he sustained to his limbs and internal organs were such that he never recovered and although he successfully sued the city's sewer department and was awarded $2,000, it was not paid before his death in May 1854.

Lewis died destitute and his wife and son were forced to sell his goods and property to cover his debts. The city only paid up three years later.

Lewis' harpoon has been described as "the single most important invention in the whole history of whaling" but, perhaps, in a small way, the whales got their revenge.

22. Garrett Morgan (1877 – 1963)

What do a gas mask and a three-light traffic light have in common?

They were both the brainchild of Garrett Morgan.

Born to former slaves, Garrett only had a rudimentary education but his innate business sense, inventiveness and strong work ethic soon saw him make a name for himself.

In 1905, working as a tailor in Cleveland, he was experimenting with a liquid to give sewing machine needles that extra sheen. He wiped his hands on a woollen cloth and was astonished the following day to see that the woolly texture of the cloth had straightened out. Fascinated, he began experimenting with other textures, including the coat of an Airedale terrier. So straight was the dog's coat after its treatment that its owner failed to recognise it.

Plucking up courage, Garrett used the liquid on his own hair and found that it straightened his curls. Sensing an opportunity, he converted the liquid into a cream, took out a patent and launched the G A Morgan Hair Refining Company to market his product. The first chemical hair straightener was born.

It made him a comfortable living. He soon added a sewing machine repair shop to his business portfolio, becoming along the way the first African American to own a car in Cleveland.

Garrett's interest in fire safety was piqued when he saw firefighters struggling for air when they entered a smoke-filled room. Recognising that smoke and fumes tend to rise leaving the more breathable air at lower levels, his smoke hood, as he called it, had a series of tubes which dangled down towards the ground. These drew the (relatively) cleaner air to a mask, which contained a wet sponge to filter out any smoke and cool the air.

Simple and effective, albeit cumbersome by modern standards, it was streets ahead of its rivals, which were either difficult to put on, complicated or simply just didn't work.

Satisfied that he had a commercial product, Garrett applied for a patent, which he received in 1912, and established a company, the National Safety Device Company, in 1914 to market it.

Inventing is one thing but selling is another and it proved particularly taxing for a man of African American origin in the still segregated South of the United States.

Nothing if not inventive, Garrett hit upon a cunning strategy. He hired an actor to pose as the inventor, while he dressed up as a native American chief, Big Chief Mason. It would be announced that Mason would enter a smoke-filled teepee for ten minutes wearing the mask. Twenty-five minutes later Mason aka Garrett would emerge from the tent, none the worse for his ordeal.

Whether these sales tactics turned into hard orders is difficult to determine as no sales figures have ever been found.

Garrett's smoke hood was instrumental in getting to the injured and saving lives in the aftermath of a tunnel explosion under Lake Erie on July 24, 1916. There had been two failed attempts to reach the victims before Garrett arrived at the scene. He donned his smoke hood and was the first to lead the charge.

But so endemic was institutionalized racism at the time that the Cleveland city officials and the newspapers wrote Garrett out of the story and his prominent role both as a heroic rescuer, and the provider of safety equipment that actually worked, was only grudgingly accepted many years later.

Perhaps that is why Garrett established his own newspaper, the *Cleveland Call*, in 1920.

Garrett persisted, made modifications to his mask, including providing it with its own air supply, and did receive recognition from some quarters, the International Association of Fire Engineers making him an honorary member and awarding him a medal.

The bitterest blow to Garrett must have been the decision of the United States Army, when they entered the First World War, to use the British Small Box Respirator and the French M2 Respirator as their gas masks of choice rather than their home-grown version.

And now to traffic lights.

There was nothing new about traffic lights, the first, the invention of a railway signalman, J P Knight, having been installed outside the Houses of Parliament in Westminster in 1868. However, the increase in the volume of traffic, human, animal and now motorized, and the relative speed at which the new-fangled cars travelled meant that there needed to be a means of regulating their flow, especially at busy cross roads.

The first generation of traffic signals consisted of two lights, stop and go, and were often manually operated. They gave little time for the driver to react when the lights were changed and there could be some spectacular crashes as a result.

The story goes that after witnessing one such accident, Garrett hit upon the idea of introducing an intermediate light which warned the oncoming motorists of a change to stop or green.

In 1923, Garrett was awarded a patent for his design of a T-shaped pole with three settings. It was not the first such, William Potts had developed the first in 1920, but a feature of Garrett's design was that it could be set at half-mast and blink at night.

He sold the patent on to General Electric for $40,000.

In his later years, Garrett was dogged with ill health and glaucoma rendered him effectively blind. But he carried on inventing, developing, amongst other things, a self-extinguishing cigarette which had a water pellet just above the filter.

I wonder what happened to that?

Even for a relatively successful African American, Garrett had to battle against institutional racism to succeed.

23. Elijah McCoy (1844 – 1929)

Who was the real McCoy?

I am a logophile and in my etymological researches I have found that there are often a number of contenders for the origin of a phrase or idiom. It requires a lot of diligence to sort the wheat from the chaff. The real McCoy means the real deal, the original article. One of the contenders for its origin is Elijah McCoy.

Born to former slaves from Kentucky, who had escaped to freedom in Canada, via the Underground Railroad, Elijah qualified as an engineer, eventually settling in Michigan.

Looking for work, he could only find a position on the Michigan Central Railroad. The railways operated a strict segregation policy and deemed that a person of colour could not possibly be skilled enough to perform the important role of an engineer. Instead, he was deployed as fireman, stoking coal into the voracious boilers of locomotives.

I am of an age to have seen and travelled on steam locomotives. The sense of power and the great plumes of smoke, environmentally unfriendly, for sure, were thrilling for a small boy and I loved standing on a bridge to be enveloped by the smoke from a train thundering by.

The railways played their part in opening up countries, facilitating the speedy transfer of goods, encouraging the development of suburbia, and spearheading the concept of leisure time and holidays for ordinary people.

One of the principal issues with steam engines was that their many moving parts needed to be oiled and lubricated on a regular basis. To do that, the early locomotives had to often stop and be serviced, impacting the reach and performance of the engines and eating into the profits of railway operators.

The first person to apply successfully their grey cells to the problem of lubricating a steam engine on the move was Englishman, John Ramsbottom, who, in 1860, came up with the displacement lubricator. It used the steam from the engine to enter a valve containing oil, pushing the oil out on to the moving parts. Adopted by the Great Western Railway, its principal problems were that you could not regulate the flow of lubricant and that it only worked when the engine had a head of steam.

Working away in a little workshop at his home in Ypsilanti in his leisure hours, Elijah investigated ways in which he could automate the lubrication of a steam engine's moving parts in a more efficient way than Ramsbottom's device.

By 1872, he had come up with what he described as a "lubricating cup", which dripped oil when and where it was required or which, as he described it more verbosely in his patent application, "provides for the continuous flow of oil on the gears and other moving parts of a machine to keep it lubricated properly and continuous and thereby do away with the necessity of shutting down the machine periodically".

The patent (US Patent 129,843, "Improvement in Lubricators for Steam-Engines") was granted in 1872.

It was well received and orders flooded in from railway operators around the States. Despite the patent, the actual device was easily replicable and with modest alterations other manufacturers were able to come up with rival lubricators. However, such was the quality and efficiency of Elijah's lubricator that train operators insisted on getting their hands on the real McCoy, or so it is claimed.

Lack of capital dogged Elijah.

He continually modified and enhanced the lubricator, making it capable of being used on a variety of other machines such as ships, oil drilling rigs and mining equipment, accumulating some fifty or so patents. To fund the work, he often had to sell his patents or at least a percentage stake in them. We will look at this practice in more detail (see Nos. 49 to 50). It was not until 1920, that he established his own company, the Elijah McCoy Manufacturing Company, following his development of a graphite inductor which allowed the latest generation of locomotives to be lubricated.

Curiously, he was barely mentioned in the literature about lubrication in the early twentieth century, being entirely written out of the pages of E L Ahron's *Lubrication of Locomotives*, published in 1922, almost certainly on the grounds of his race.

In 1922, Elijah and his wife, Mary, were injured in a serious car accident, Mary fatally, and from then on until his death in 1929, Elijah was dogged with financial, physical, and mental problems.

And was he the real McCoy?

I'm not sure. A variant of the phrase, with an identical meaning, appeared in a Scottish poem, *Deil's Hallowe'en*, dating to 1856; "a drappie o' the real McKay", McKay being a whisky. The phrase appeared frequently in Scottish newspapers in the 1860s. Elijah may have been a worthy substitute, given the quality of his lubricators, but cannot have been the reason why the phrase came about in the first place.

24. Thomas L Jennings (1791 – 1859)

Is there anything more annoying than spilling something down your clothes?

On reflection, plenty but perhaps because my manual dexterity isn't quite what it used to be, it seems to happen to me with increasing regularity. I'm forever dabbing and rubbing items of clothing, trying to get a food stain out. Depending on the combination of foodstuff and fabric, some stains seem immoveable and the only thing for it is a trip to the dry cleaners.

On such a trip, I pondered; someone must have invented dry cleaning. Who was it?

Step forward, Thomas Jennings.

Born in 1791, a free African American, this was to be an important distinction as his story unfolds, Thomas became an accomplished tailor. So good were his skills that people came from miles around either to have their clothes altered or bespoke apparel made by him. Soon he had amassed sufficient funds to open his own store on New York's Chapel Street.

People were no less clumsy with their drinks and food then than they are today, but they didn't have the option of popping down to the dry cleaners.

Their choice was either to get the stain out as best they could and continue wearing the garment or to consign it to the bin, an expensive option. Whilst replacing soiled clothing with new was grist to the mill of Thomas' tailoring business, he didn't like to see garments that he had worked on for hours on end discarded before they had reached the end of their natural life.

So, Thomas began to experiment on ways to clean clothing, deploying a range of different solutions and cleaning agents on a wide array of fabrics. Eventually, he hit on a winning combination. After extensive testing, in 1820, he applied for a patent for a process he called "dry scouring", the forerunner of dry cleaning.

Thomas was awarded a patent (US Patent 3306x) on March 3, 1821, making him the first African American to hold one.

As we saw with Benjamin T Montgomery (see No 19), the law at the time did not perceive slaves as citizens of the United States and so they were unable to swear the oath necessary to stake their claim to their invention. This legal impediment did not impact Thomas. He was a free man, after all, and he was able to benefit from his ingenuity.

And benefit he did.

Dry scouring became an extremely popular way to clean badly soiled clothing and Thomas made his fortune. But, alas, we know very little about the particulars of the method Thomas had developed because the US Patent office was destroyed by fire in 1836 and its records went with it. In 1825, Jolly Belin opened what is thought to have been the first commercial dry cleaning laundry, in Paris, using turpentine.

Perhaps this was Thomas' process.

The problem with turpentine was that it made the clothes smell but it was not until the 1850s that petroleum-based substances were used to dry clean clothes. These substances were highly inflammable and there were often by-laws in place prohibiting dry cleaning premises to operate in densely populated areas. Clothes were often brought into a shop in the town and then sent to a laundry out in the sticks to be cleaned.

Less dangerous chlorinated solvents were only used after the First World War.

Thomas used his wealth to purchase the freedom of those relatives of his who were still enslaved and then fund the abolitionist movement in the North-Eastern states, becoming, in 1831, assistant secretary for the First Annual Convention of the People of Colour, held in Philadelphia.

His daughter, Elizabeth, was a chip of the old block.

She was an activist on behalf of the abolitionist movement, like her father, and one day whilst on a New York City streetcar on her way to church she was ordered off. She sued the operators, Third Avenue Railroad Company, on the grounds of discrimination and in 1855 the Brooklyn City Circuit found in her favour.

The very next day, the company desegregated their buses. Her attorney, Charles Arthur, went on to become US President in 1881.

25. Dr. Charles Drew (1904 – 1950)

I wouldn't say I was haemophobic but I'm rather squeamish. I would prefer it if my blood and the blood of others stayed where it should be, pumping around the arteries and veins of our bodies. That said, I have enormous respect for those individuals who generously give of their blood to assist others in emergencies or to be stored in blood banks.

But who came up with the concept of blood banks?

Charles Drew was born in Washington, D.C, attended medical school at the McGill University in Canada, where his interest in blood storage was piqued, graduating in 1933. While earning a doctorate at Columbia University in 1938, he won a scholarship to study under the eminent surgeon, Allen Whipple, at the Presbyterian Hospital in New York.

But instead of taking the traditional path of surgical life, learning about surgical pathology, operating procedures and the like, which would have involved having direct access to patients, Charles, an African American, was assigned to work with John Scudder, who had just secured funding to establish an experimental blood bank.

A piece of institutional racism which would redound to society's benefit.

Charles set to his work with gusto, reviewing and assessing previous research literature on blood and transfusion methodologies, blood chemistry and fluid replacement and looked at the variables which affected the shelf-life of stored blood. Nothing escaped his attention as he investigated the impact of different types and amounts of anticoagulants in the blood and preservatives and even the shape of storage containers and the impact of ambient temperature.

His thesis, entitled *Banked Blood: A Study in Blood Preservation*, described by Scudder as "a masterpiece" and "one of the most distinguished essays ever written, both in form and content", earned Charles a doctorate, the first African American to receive a medical doctorate from Columbia.

Even Whipple recognised his talents.

The principal finding from Charles' researches centred around plasma, a clear, yellow liquid full of proteins and electrolytes, which carries blood cells around the body. Plasma was a blood substitute and Charles found that there were advantages to using it instead of whole blood in emergencies, principally because it kept longer without refrigeration, could be used with any blood type, was less likely to transmit disease, could be injected, in large amounts, through veins, muscles and the skin and, critically, it would not deteriorate if shaken about when transported.

The timing of Charles' researches couldn't have been better.

The Second World War had broken out, the British were incurring casualties and there was an urgent need to get blood, or an equivalent, out to the troops to save lives. Charles was asked to direct a programme to organise the collection, testing and distribution of blood plasma in Britain.

Of course, the trick was to work out how to separate the plasma from blood cells.

Charles and Scudder developed a technique involving the use of centrifuge and sedimentation. After a complex procedure, the batch of plasma was diluted with a sterile saline solution and then tested for bacteria before being sealed in a container and packed for shipping. The first batch received in Britain was confirmed to be "entirely satisfactory".

During the five months the programme lasted, between late 1940 and early 1941, Charles had collected blood from 15,000 people and given around 1,500 transfusions.

Having established the success of the programme and his methodology, Charles was appointed the first director of the American Red Cross Plasma Bank. During the course of America's involvement in the Second World War, Charles' organisation had recruited 100,000 donors, saving the lives of thousands of servicemen. One of his innovations was the bloodmobile, a blood donation lorry with an in-built refrigerator.

But the dead hand of institutionalised racism raised its ugly head once more.

The US army enforced a strictly segregated approach to blood transfusion. Initially, non-whites could not give blood. The policy was then modified to the extent that blood donated by non-whites could not be given to whites. Charles denounced the policy as "unscientific and insulting to African Americans". He was asked to resign his position, which he did.

After the war, he became Chief Surgeon at the Freedman's Hospital in Washington.

On April 1, 1950 disaster struck. On his way to a medical conference at the Tuskagee Institute in Alabama, after a night shift, Charles fell asleep at the wheel of his car, his foot got caught under the brake pedal and the car plunged off the road. Within half an hour of arriving at the Alamance Hospital, he was pronounced dead. Stories quickly circulated that white doctors had refused to treat him, but these are probably false. The severity of his injuries and the loss of blood were such that his demise was inevitable. Indeed, according to some, a blood transfusion would not have helped.

There is some irony in that, for sure.

Part Three

Imitation and neglected sparks

THAT'S MY IDEA

IMITATION IS THE sincerest form of flattery – Charles Caleb Colton, *Lacon: or many things in few words,* 1820.

It is particularly galling for an inventor to find that someone else has stolen your glory. It may be flattering to know that there is some merit in your idea but that doesn't put bread on your table. Worse still, history relegates you to a footnote, at best, in its pages.

Well, it is time to redress the balance and bring to your attention the stories of eight individuals who, in their different ways, saw their ideas usurped by others and lost out on the glory they deserved.

26. Rasmus Malling Hansen (1835 – 1890)

Recently I had cause to write something with a pen. It was an odd feeling and the result was something that an inebriated arachnid would feel proud of. So embedded in our daily life is the keyboard in all its forms that many fear for the future of cursive script. But the typewriter, which is the granddaddy of the myriad keyboards we use, was a relatively modern innovation.

And who could claim to be its inventor?

Rasmus Malling Hansen, a Danish priest, who was principal of the Royal Institute of Deaf-mutes in Copenhagen, that's who. His work piqued an interest in the make-up of Danish letters and sounds and, in particular, the speed at which letters could be spoken and written per second. He concluded, according to his patent application, that "in a given time one can say five times as many sounds as they can be written".

This observation led Rasmus to consider how he could speed up the production of letters on paper.

From around 1865, he started experimenting and, by 1870, was sufficiently satisfied with the results to apply for, and receive, a patent for what he described as an "apparatus for quick writing". He claimed that "the writing speed will easily be two to three times as fast as normal, and practice in using the apparatus should be able to bring this speed up to speech speed".

It was an ingenious affair.

The heart of the contraption was a large brass hemisphere in which fifty-two keys were arranged such that the keys representing the most frequently used letters were directly in reach of the user's strongest and fastest typing fingers. The vowels were arranged on the left-hand side and the consonants on the right. This arrangement, together with the use of short pistons, which went through the hemisphere, enabled the user, after some practice, to reach prodigious speeds.

To the eye it looked like an over-sized pin cushion.

The paper was attached to a cylinder which could move both vertically and horizontally. Using an electro-magnet, powered by a ten or twelve-cell battery, a mechanical escapement moved the carriage the required distance each time a key was depressed. As well as the forerunner of the typewriter per se, Rasmus' machine could be claimed to be the first electric typewriter.

The machine caused a sensation when it was exhibited at the large industrial exhibition in Copenhagen in 1872, winning Rasmus first prize. It was also well received at the world exhibitions in Vienna in 1873 and in Paris in 1878.

But Rasmus was not satisfied and soon replaced the cylinder with a flat carriage, to which the paper was attached, and, in 1875, he was able to dispense with the battery, having found a mechanical solution to the problem of moving the paper.

The philosopher, Nietzsche, bought one but never got on with it.

Leaving the philosopher's lack of manual dexterity to one side, the principal problem was that the machine, undoubtedly efficient as it was, was fiendishly expensive to produce, making it difficult to attract a manufacturer to produce it in sufficient quantities to make a dent in the market. This opened the way for a rival typewriter, less efficient but cheaper and better marketed, the Remington typewriter which was first produced commercially on March 1, 1873 in Ilion, New York.

And the rest is history.

Rasmus could never interest manufacturers in his machine and when he died, his machine died with him. Very few survive but if you have one, they are worth a lot of money.

In 1872, Rasmus also invented a high-speed machine for stenography, the Takyagraf, and was the first to exploit the potentials of carbon paper, developing a technique called Xerografi, but perhaps he is best known for inventing the typewriter and failing to exploit it commercially.

27. Hans Lippershey (1570 – 1619)

"Twinkle, twinkle, little star/ How I wonder what you are" goes the nursery rhyme. There is something mystifying and deeply captivating about the celestial bodies that sparkle and shine above our heads at night-time. From time immemorial, Homo sapiens has wanted to get to know them better. Today, of course, we can get a closer view of them from terra firma by using a telescope.

But who invented this very useful scientific instrument?

Popular theory gives the credit to Galileo Galilei but, inevitably, it is a lot more complicated than that. This is where Hans, or Johann, Lippershey, a German-Dutch spectacle maker, comes in.

The techniques for making glass and grinding lenses came on leaps and bounds in the sixteenth century, making it easier to develop smaller and more powerful lenses. This prompted a revolution in the way that we saw the world, bringing distant objects closer and magnifying objects almost invisible to the naked eye to a size that we could easily see.

Naturally, as a spectacle maker, Lippershey would have some of these new-fangled lenses in his workshop. There is a story, probably apocryphal, that Lippershey conceived his idea of a telescope when two children held up a couple of lenses and made the weather vane of the local church appear closer.

Less charitable souls claim that he stole the idea from a neighbour, fellow eyeglass maker, Zacharias Jansen.

The truth is buried in the mists of time but what is certain is that Lippershey developed a rudimentary form of telescope, consisting of a concave eyepiece, which was aligned to a convex objective lens. It boasted a magnification power of three, pretty feeble by modern standards, but at least it was a start.

Emboldened by his success, on October 2, 1608, Lippershey applied to the States General of the Netherlands for a patent for what he called an instrument "for seeing things far away as if they were nearby", a rather clumsy description but the word, telescope, was only coined three years later, by Giovanni Demisiani.

Lippershey did not get a patent granted, perhaps the waters had been muddied by the controversy as to how he got the idea. Another complication was that a few weeks later the Dutch authorities received an application for a patent for a similar instrument, this time from Jacob Metius, another Dutch instrument-maker.

The emergence of a rival instrument led the authorities to draw the inevitable conclusion that the device was easy to make and, therefore, difficult to patent.

At least Lippershey received a large fee from the Dutch government in return for the use of his design. Poor Metius had to make do with a small reward.

The device created a bit of a stir and was mentioned in a report issued and distributed around Europe of the visit of the embassy of the King of Siam to the court of the Dutch crown prince, Maurice, in the Hague. The genie was out of the bottle and a number of eminent scientists began experimenting with the concept of using a pair of lenses to bring the image of something nearer to the viewer.

By the summer of 1609, the English scientist, Thomas Harriott, had produced a telescope with a magnification factor of six. He pointed his telescope at the moon and, in August 1609, drew what he saw but never published the results.

And then Galileo got in on the act.

His considerable intellect was piqued by reports of the Dutch perspective glasses, which reached him in 1609. Within days he had created his own telescope, without seeing a Dutch version, which boasted a magnification power of twenty. With this he observed the moon, discovered the rings of Saturn and four of Jupiter's moons. Galileo reproduced what he saw in a series of astonishing ink drawings, which were published.

So, while Harriott drew the moon first and Lippershey can rightly claim to have been the first to develop a telescope, Galileo scooped the glory.

Such is the fickle finger of fate.

28. Dietrich Nikolaus Winkel (1777 – 1826)

For a musician, time is important. You can dispense with melody or harmony but if rhythm goes out of the window, then you are left with an unholy racket. That is why you see a conductor flailing their arms in front of a concert orchestra or a drummer in a rock or jazz ensemble preparing a solid foundation upon which the other players can build.

When musicians are practising, they will often use a metronome, a handy device which you can set to register so many beats per minute by way of audible clicks or ticks. Being mechanical it is unerring. Composers mark their scores with metronome settings to give the musos a clue as to the tempo at which to play the piece.

Of course, some bright spark must have come up with the idea of a musical metronome and this is where Lippstadt-born Dietrich Winkel comes in.

He was not the first to develop a metronome, this honour goes to the Andalusian polymath, Abbas ibn Firnas (810 – 897 CE), who is said to have devised "some sort of metronome".

In 1696, Frenchman, Etienne Loulie, created the first mechanical metronome, using an adjustable pendulum. The problem with Loulie's invention was that it did not make a sound and did not have a device, the technical term is an escapement, to keep the pendulum in motion.

For a musician, it was not much use.

Winkel, who by 1812 had now settled in Amsterdam, began experimenting with pendulums. His breakthrough came when he realised that by weighting a pendulum on both sides of a pivot, it could beat a regular rhythm, which was audible. It could be adapted to suit various tempi and was housed in the now familiar pyramid casing.

Winkel donated his "musical chronometer" to the Hollandsche Instituut van Wetenschappen on November 27, 1814. It was described and commended in the *Journal of the Netherland Academy of Sciences* the following year.

If Winkel thought by developing this machine he was on to a winner, he was gravely mistaken. He made the fatal mistake of failing to patent his musical metronome. This opened the way for Johann Nepomuk Maelzel. He tried to buy the rights and title to Winkel's metronome. When Winkel refused, Maelzel simply copied his machine, added a scale and applied, successfully, for a patent.

Maelzel produced around 200 of his metronomes and sent them out to friends, composers and manufacturers of musical instruments for their comments and suggestions for modifications. One recipient was Ludwig van Beethoven, who was much taken by the device and added metronome settings in his later scores.

Winkel sued Maelzel and won but by then the damage had been done. Although the courts acknowledged our hero as the true inventor of the metronome, Maelzel had cornered the market. Even to this day the metronome is known as the Maelzel Metronome and the notation MM is used in score to denote the tempo at which a piece is to be played.

Winkel did achieve some fame of sorts by inventing the componium, an organ with two barrels which revolved automatically. The barrels took turns at playing a variation of a piece whilst the other randomly, by way of something resembling a roulette wheel, selected the next variation to play. The variations were almost limitless and it could play variations, "not only during years and ages, but during so immense a series of ages that though figures might be brought to express them, common language could not".

It wowed the crowds when it was displayed at an exposition in Paris in 1824.

But for Winkel, the metronome was the one that got away.

29. Richard Pearse (1877 – 1953)

If you are an inventor, there is no point in hiding your light under a bushel. Take the case of Richard Pearse. It didn't help that he was reclusive by nature and lived almost at the end of the world in New Zealand's South Island.

Richard, a farmer, was a serial inventor with a thing for bamboo. In 1902, he patented his first invention, a bamboo-framed bicycle powered by a vertical-drive pedal action and with a rod and rack gear system, integral tyre pumps and back-pedal rim-brakes. It was certainly ingenious, the like of which had not been seen before.

But like many a chap with an inventive streak at the time, Richard's imagination was piqued by the early attempts to achieve powered flight.

It seems he had been mulling over the challenges in his mind from as early as 1899 and, by 1902, had come up with the design for a petrol engine. Naturally, it was ingenious, consisting of two cylinders with pistons. The clever bit was that the pistons worked in either direction, effectively turning into a four-cylinder engine, capable of generating between fifteen and twenty-two horse power.

Then Pearse built a plane to house the engine out of bamboo, naturally, tubular steel, wire and canvas. It was the first to use proper ailerons and had a modern-style tricycle landing system, which meant that it could take-off and land anywhere. It was also a monoplane.

In comparison with the Wright's prototypes at the time, it was superior save for a rather crude propeller system.

Having built his plane, the big question was; would it get off the ground?

Chronology is far from certain, not least because Pearse was reticent to publicise his experiments, which were often conducted in secrecy on his own extensive land. What eye-witness reports there are seem somewhat vague on key points of his exploits.

But it seems that as early as March 3, 1903, Pearse took to the air in a public demonstration of his flying machine. Pearse manoeuvred his machine on to the Main Waitohi Road, which ran along the perimeter of his farm. The plane took off, perhaps reaching a height of some four metres, and travelled in a straight line for between fifty and

400 metres, there is no consensus amongst reports, before coming to rest in a gorse bush.

Some accounts date this attempt at flight as early as the end of March 1902.

Undaunted, Pearse continued with his experiments. On May 11, 1903 he took off along the side of the Opihi River, turning left to clear the thirty-foot high river bank, and then manoeuvring to the right to follow the line of the river for some 1,000 yards. It was at this point that his engine overheated and began to lose power, forcing Richard to land in the riverbed. A local, Arthur Tozer, was crossing the river bed in a horse-drawn carriage at the time and was astonished to see Pearse fly directly over his head.

As well he might!

Pearse never publicised his achievements and the local newspapers only picked up on the story in 1909, perhaps setting a world record for the slowest pack of newshounds.

Pearse himself sowed confusion in the chronology by claiming in a couple of letters, published in 1915 and 1928, that it was not until February or March 1904 that he "set out to solve the problem of aerial navigation".

I know as I get older memory plays tricks on my grasp on chronology but there are enough independent witnesses to suggest that Pearse was experimenting with a version of manned flight, it was hardly controlled, far earlier than that.

Pearse patented his aircraft in 1906 but then seemed to lose his interest in flight, frustrated that competitors from abroad were getting all the glory. In the 1930s, though, he did design, and partly build, an aircraft with foldable wings and tail, which could be stored in a garage.

The idea never took off.

For the record, the Wright brothers flew the Kitty Hawk on December 17, 1903, more than half a year after Pearse.

30. Konstantin Tsiolkovsky (1857-1935)

I'm fairly catholic in my reading but there is one genre that I can't really get on with, science fiction. Perhaps it is my lack of imagination or just that I would prefer to spend my time understanding the range of emotions that make us humans tick or how we react to situations, comic or tragic.

I'm sure it is my loss.

But there are some whose imagination is stimulated by sci-fi and one such was Konstantin Tsiolkovsky.

The fifth child out of eighteen born to an impoverished Polish immigrant family in Russia, (cause and effect, I can't help thinking), profoundly deaf after a childhood bout of scarlet fever, and pretty much self-taught, Konstantin stumbled upon Jules Verne's *From the Earth to the Moon*, first published in 1865.

Fascinated by the prospect of travel to the Earth's nearest neighbour but being of a practical bent, he calculated that using a giant cannon to fire a spacecraft to the moon, Verne's designated method, would generate forces that would kill the unfortunate passengers.

Verne did, though, light the blue touch paper that ignited Konstantin's life-long interest in all matters aeronautical. He is reported to have remarked, "I do not remember how it got into my head to make the first calculations relating to the rocket. It seems to me the first seeds were planted by the famous fantasoeur, J Verne".

Initially, he set his sights on flight, designing early airships and Russia's first wind tunnel. He published his first work on the subject in 1892. In 1894, he wrote an article in which he proposed an aircraft made of metal. Surely the idea would never take off.

But the lure of space travel proved too great.

Konstantin tried his hand at writing science fiction but found that his mind wandered to trying to solve the practicalities of getting a rocket out of the Earth's atmosphere and on its way to the moon. From 1895, this became his major preoccupation.

By 1903, Konstantin had cracked the problem, writing *Explorations of the World Space with Reaction Machines,* which was published in Russia's scientific review, *Nauchnoe Obozrenie.* More articles were forthcoming from the prolific scientist. His rockets were to be fuelled by a mix of liquid hydrogen and liquid oxygen, precisely the same mix as was to be used by the Space Shuttle.

Astonishingly, hydrogen had only been liquefied for the first time by James Dewar in 1898.

Konstantin developed what later became to be known as the Tsiolkovsky Equation, which demonstrated the mathematical relationship between the change in the mass of a rocket as it burnt fuel, the speed of its exhaust gases, and the final velocity of the rocket. It became the bedrock that enabled the later development of astronautics.

Konstantin wasn't done.

In 1929, he published an article in which he postulated that in order for a rocket to break out of orbit, it would need a series of rockets to drive it forward, each one breaking off from the main body of the craft as it had used up all of its fuel.

Who needed science fiction when you had Tsiolkovsky?

But hardly anyone outside of Russia had heard of his work. The Bolshevik revolution meant that very little hard information was coming out of the country. In any event, Konstantin was a lowly school teacher, who spent his spare time thinking about rocketry, rather than a fully-fledged scientist attached to an acknowledged academic institution of standing. Moreover, the scientific journal he used to publish his articles had been closed down.

There was no world-wide web to publicise his findings.

So, independently and in parallel during the 1920s, the German Hermann Oberthand, and the American physicist, Robert Goddard, worked on many of the problems that had exercised Konstantin's mind and often came up with the same conclusions as he had. All three could claim to be the fathers of rocketry, although Konstantin seemed to have got there first.

Full recognition of his genius only came posthumously. His work was drawn on and influenced the rocket designers, Valentin Glushko and Sergey Korolyov, as Russia strove to win the space race in the 1950s and early 1960s. The most prominent crater on the dark side of the moon bears Konstantin's name as does asteroid 1590.

Tsiolkovsky was a great visionary. He wrote that "mankind will not forever remain on Earth, but in the pursuit of light and space will first timidly emerge from the bounds of the atmosphere, and then advance until he has conquered the whole of circumsolar space."

He was not wrong.

31. Geoffrey Dummer (1909 – 2002)

If you want a graphic illustration of the difference in the appetite for risk between state organisations and private industry, then you could do worse than look at the tale of Geoffrey Dummer.

These days we are used to, and indeed expect, electrical gadgetry to pack as much punch and functionality into as small a space as possible. What has transformed the size, complexity and sophistication of these devices has been the adoption of the integrated circuit board. Open one up and you will find a little box of tricks, a compact miniature board, often printed these days, which brings together a range of components with different functionality into one discrete space. It may not be too fanciful to suggest that the integrated circuit board has had as much an impact on the development of the digital age as steam power had for the Industrial Revolution.

The person who came up with the idea of integrated circuitry, the forerunner of the internal circuit board, was a British scientist by the name of Geoffrey Dummer.

Dummer spent much of the Second World War designing, developing, and installing test equipment upon which to train radar operators at the Telecommunications Research Establishment in the beautiful Worcestershire town of Malvern. Fundamental to improving the performance and efficiency of the training equipment was eliminating the number of electrical connections between components, which were often unreliable, particularly in the challenging conditions to be experienced in wartime.

His wartime work led Dummer to conclude that it should be possible to put together a number of circuit elements on to something like silicon. Once peace had returned, he started to apply his grey cells to the problem.

By 1952, Dummer was ready to unveil his findings. On May 7, at the Symposium on Progress on Quality Electronic Components, addressing the audience, he said, "with the advent of the transistor and the work in semiconductors generally, it seems now possible to envisage electronic equipment in a solid block with no connecting wires." He went on, "The block may consist of layers of insulating, conducting, rectifying and amplifying materials, the electronic functions being connected directly by cutting out areas of the various layers."

In other words, what he was describing is an integrated circuit board.

Having the vision is one thing, bringing it to fruition is another.

By this time Dummer was working for the Ministry of Defence and, whilst he had a free hand to try out his ideas, he found it difficult to put together a fully functioning board. As is often the way with government departments, particularly in a country which was painfully pulling itself free from the debts incurred during the Second World War, the bean counters lost patience and pulled the plug on Dummer's funding.

In Dummer's own words, "nobody would take the risk. The Ministry wouldn't place a contract because they hadn't an application. The applications people wouldn't say we want it. It was a chicken and egg situation."

Now Jack Kilby, a scientist employed by Texas Instruments in the United States, enters our story.

Recognising that the principal elements of a successfully functioning integrated circuit board were fault-free components and as few wires as possible in the interconnections to enhance the speed of the circuit, Kilby's idea was to build a single device out of silicon, which would then be soldered on to a circuit board. By eliminating the wires between parts, he could squeeze more parts into a smaller space.

Having built his prototype, Kilby showed it to his boss and was given the green light to file for a patent on February 6, 1959 for "a body of semiconductor material… wherein all the components of the electronic circuit are completely integrated."

Meanwhile, over in California, Robert Noyce at Fairchild Electronics was working, independently, on the concept of putting an integrated circuit on to one chip. During 1960, realising that Texas Instruments had filed a patent application for their version of the circuit using Kilby's design, Noyce filed a very detailed application for his version. It took him six months to complete, hoping that it would not be adjudged to infringe Kirby's copyright.

Despite submitting his application some five months after Kilby, Noyce received his on April 25, 1961 (US patent No 2,981,877). Kilby's wasn't granted until June 1964 (US patent No 3,138,743).

Texas Instruments and Fairchild then went to court to determine who had the rights to the board. A compromise was struck in 1966 when both companies agreed to licence the product to each other. However, anyone else wanting to use it had to pay a fee to both companies. Both Kilby and Noyce were recognised as inventors of the integrated circuit but the Nobel Prize Committee only got round to recognising the achievement in 2000. By that time, Noyce had died, in 1990, and as Nobel Prizes are not awarded posthumously, only Kilby was cited in the award of the Nobel Prize in Physics for that year.

As for poor Dummer, although he was still alive in 2000, his contribution to the story had been forgotten. In peacetime, it seems, American companies are more willing to invest in research and development than the British government. Had that not been the case, Dummer would have received his just desserts. His contribution to the story is only now being recognised.

32. Angela Ruiz Robles (1895 – 1975)

Books do furnish a room. You can tell a lot about a person by the presence or absence of books in their house. When I encounter a bookshelf, I feel drawn towards it, as if I am answering the siren call. There is something magical about the physical properties of a book, the feel, its weight, the cover, the spine, its illustrations, the layout of the text, even the type selected.

Beautiful as they undoubtedly are, they are heavy and take up a lot of room.

I'm a voracious reader and get through books by the dozen. I have a few favourites, which I return to from time-to-time, but most of my reading matter is engorged once and once only. And one of my personal nightmares is being away from home, travelling or on holiday and running out of reading material.

To me and, I'm sure, many others, the e-reader is a Godsend, allowing me to have almost instantaneous access to hundreds of books in a portable rectangular device. Aesthetically pleasing it is not and unlikely to revolutionise the way books are delivered as the format's early evangelists once claimed, but it is convenient and, for bookworms like me, an invaluable support prop.

The concept of an automated reading device dates back to the 1940s, the brainchild of the director of the Instituto Ibanez Martin in Ferrol in Spain, Angela Ruiz Robles. Her vision was to make teaching easier and to enable her students to maximise their knowledge with the minimum of effort.

Fundamental to achieving this aim would be the development of a mechanical book, which contained all the texts that a student would need. Instead of volumes of battered text books, all their satchels would contain would be a light-weight, portable, easy-to-use mechanical reader.

Angela worked away on her idea and by 1949 had come up with a pastel-green coloured metal box which she called, snappily, I feel, Procedimiento mecánico, eléctrico y a presión de aire para lectura de libros or, in English translation, "a mechanical, electrical and air pressure procedure for reading books".

Inside were a series of tapes on interchangeable spools, some containing text and others illustrations, all protected by a transparent and unbreakable sheet. It came with a magnifying lens and a light so that it could be used in the dark. The mechanical encyclopaedia even had an audio component, which brought the text to life.

Angela had considered a wider application for her book than just Spain, proposing alphabets and texts in a number of languages. Content could be read from start to finish or the reader could skip to a new chapter by pressing a button. She even envisaged an interactive index and a list of installed works, which the student could move between by pressing one or more buttons.

To entice the publishers, Angela proposed a standard size for cartridges and, of course, some of the production costs associated with book production, such as pasting and binding, would be eliminated.

What was there not to like?

Satisfied with her prototype, Angela applied for a patent. On December 7, 1949 she was awarded Spanish patent 190,698 for what was described as a mechanical encyclopaedia. She paid the annual renewal fee up until 1961 but was unable to attract sufficient funding or interest from publishers to make her vision of an alternative to a book a commercial reality.

Undaunted, on April 10, 1962, Angela applied for and received a patent (No 276,346) for an "apparatus for diverse readings and exercises". Although it contained many of the components of the original mechanical encyclopaedia, it had a slightly more streamlined design. Be that as it may, it still met the same fate as Angela's original machine. No manufacturers or publishers would back it with cash to bring it into production.

And, so, the idea of a mechanised book or reader, as we would now call it, withered and died, only to be picked up again by Michael Hart in 1971 with the prototype of a truly electronic reader.

Belatedly, Angela's contribution to the development of the e-reader has begun to be recognised but she missed out on the commercial gains of her brainwave. A version of her early prototype, a splendid affair made from bronze, wood, zinc, and paper, can be seen to this day at the Science and Technology Museum of La Coruna.

33. Elizabeth J Magie (1866 – 1948)

I've always had a love hate relationship with that board game that is trotted out when families and friends gather, Monopoly. On the one hand, it is enjoyable, engaging and can keep everyone entertained for several hours. On the other hand, for someone with socialist leanings, it disturbs me that it seems to bring out the worst features of a grasping capitalist out in many of the game's participants.

Elizabeth Magie, known to her friends and family as Lizzie, was from Scottish immigrant stock, living in Prince George's county in Washington D.C at the turn of the 20th century. She was known for her progressive political views and was looking for a way to bring her concerns about the economic impact of the monopoly on land and property owners on the common folk to a wider audience. At the time board games were becoming increasingly popular amongst middle class families and this seemed to be the best medium to spread her message.

What Lizzie developed was a game called the Landlord's Game, designed, as she said, as "a practical demonstration of the present system of land-grabbing with all its usual outcomes and consequences...contain[ing] all the elements of success and failure in the real world".

The elements of the game will be familiar to many readers, players progressed around the outer rim of a board, receiving $100 every time they passed the Mother Earth space and going to jail if they trespassed on land. Properties were available to buy and then collect rents from. Those unfortunate enough to run out of money were sent to the Poor House. There were two sets of rules; one which rewarded all players when wealth was created and one where the goal was to create monopolies and crush opponents.

Satisfied with her game, Lizzie applied for and was granted a US Patent (no 748,626) on January 5, 1904. The Landlord's Game gained some popularity with intellectuals and university campuses (the two are often mutually exclusive, I find) and was revised and improved over time. In 1924, recognising that her patent had expired and that she needed to re-establish her ownership of the game, Lizzie applied for and was granted another patent (no 1,509,312).

It is now time for Charles Darrow to enter our story.

In late 1932, Charles, unemployed and desperate for money, was introduced to a property board game by Philadelphia businessman, Charles Todd. Darrow was taken by the game and saw an opportunity to make some money, initially hand-producing the game which he called Monopoly and then a printed version, obtaining a copyright for it in 1933.

Sales of the game were so promising in the run up to Christmas 1934 that Parker Brothers, now part of Hasbro, approached Darrow and, on March 18, 1935, bought the game, the remains of Darrow's stock and helped him to secure a patent. But a month later Parker Brothers became suspicious of Darrow's claims that he was the sole inventor of the game and, in a smart move, approached Lizzie to buy the patent for her Landlord's game and a couple of other games she had created. They proposed to her a payment of $500, which she accepted, but they did not offer her a share of the royalties.

At the time, Lizzie didn't smell a rat, even writing to the treasurer of Parker Brothers that when the prototype of her game arrived, "she had a song in [her] heart". But in January 1936 her mood changed. The *Washington Evening Star* carried a picture of her holding a board from her Landlord's Game and one from a game called Monopoly. The similarities were striking, as they would be. She was angry, steadfast in her belief that Parker Brothers had stolen her best-seller of an idea.

A best seller it was too, taking off in the States and becoming an international favourite. Darrow, who had a slice of the royalties, made millions from the game. When asked by the *Germantown Bulletin* how he came to invent such a wildfire success, Darrow replied, somewhat ingenuously, "it's a freak…entirely unexpected and illogical".

Lizzie's role in developing the game was effectively airbrushed out of history. When she died in 1948, a widow with no children, neither her obituary nor her tombstone bore any reference to her role in developing one of the world's best-known games. The website of Hasbro, ironically named for the eighth year in succession in 2019 as one of the world's most ethical companies by the Ethisphere Institute, is silent on her contribution.

There matters would have remained but for an American economics professor from San Francisco University, Ralph Anspach. In 1974 he launched a game called Anti-Monopoly and was immediately sued by Parker Brothers for breach of copyright. In preparing his defence, Anspach uncovered the history of the Landlord's Game, Lizzie's role in developing it, and how Darrow had been economical with the actualité in explaining how he had come across Monopoly. After a ten-year legal battle, Anspach prevailed and Lizzie has now begun to receive the credit she deserved, at least in some circles.

NOBODY LISTENS TO ME

A SPARK NEGLECTED makes a mighty fire – Robert Welch Herrick, *Hesperides: Works both human and divine,* 1648.

One of the problems an inventor can face is that their ideas are just so out of line with received wisdom that they find it hard to get them taken seriously. Some inventors are made of sterner stuff and battle against all the odds to get their ideas accepted, no matter what the personal costs are. Others, though, simply give up, leaving the field clear for others to exploit.

Here are the tales of five individuals who found it hard to get a hearing. But once their ideas were out in the open, to paraphrase Herrick, a mighty fire broke out.

34. Ignaz Semmelweis (1818 – 1865)

Sometimes you discover something and can't persuade the powers that be that you have made a major breakthrough. This was the fate that befell the Hungarian obstetrician, Ignaz Semmelweis.

Ignaz studied Law at the University of Vienna in 1837 but switched to medicine the following year and after gaining his doctorate in 1844, he decided to specialise in obstetrics. He took up his first appointment in 1846, as an assistant in the Vienna General Hospital's maternity ward.

There were two Clinics, A, which was the preserve of doctors and trainees, and B, which was staffed by midwives only. In the mid-nineteenth century giving birth was a precarious business, often proving fatal to either the mother or the baby or, in some cases, both.

Clinic A had a phenomenally high mortality rate, about 10%, mainly because patients contracted puerperal fever, whereas the mortality rate in Clinic B was a still shocking, but lower, 2%.

Women who came to the hospital, they were mainly from the lower classes, tried as best they could to avoid Clinic A because of its fearsome reputation. Many preferred to give birth in the streets, where the mortality rate was considerably lower.

Why was that, Semmelweis wondered?

The duties of the doctors at the hospital were many and varied. They would routinely examine diseased corpses in the mortuary, carrying out autopsies to determine cause of death or dissections to further their knowledge of the human anatomy, before moving on to the maternity ward. Whilst we now tend to regard, or at least hope, that medics are as close to the Platonic paradigm of cleanliness as is possible for mere mortals to reach, in Semmelweis' time it was rare for a medic to wash their hands between dealing with patients.

He noted the discrepancy between mortality rates where doctors were involved and where midwives, who did not handle dead bodies, were in attendance. Sammelweis concluded that some form of cadaverous material, picked up from the corpses, was contributing to the high incidence of puerperal fever.

Acting upon these observations and hypotheses, Ignaz decided that he and his colleagues should wash their hands in a solution of chlorinated lime, principally to remove the smell of putrefying flesh, after handling dead bodies. The results were astonishing, with fatality rates plummeting from around 18% to 2.2% and, even for a couple of months, down to zero.

At the time, received medical opinion was that disease was spread by what was known as miasma, a poisonous vapour or mist, foul-smelling and composed of particles from decomposed matters. Not everyone accepted this theory, though. In 1762, the Austrian physician, Marcus Antonius von Plenciz, published a book, *Opera medico-physica*, in which he drew a distinction between contagious diseases which were epidemic and those which were not and suggested that specific microscopic organisms, he called them animalcules, were responsible for specific diseases. Plenciz's ideas were rejected by the general medical community.

Whilst rudimentary germ theory was certainly in the air at the time and it is possible that he was aware of it, nonetheless, Sammelweis was unable to give a rational explanation as to why washing hands had such a dramatic effect on death rates. Undeterred and convinced that he was on to something, Semmelweis began to promulgate his views. This led to great outburst of hand-wringing, but not hand-washing, amongst the medical profession, many of whom were outraged at the suggestion that their hands could be unclean.

They were gentlemen, after all.

In revolutionary Vienna, Semmelweis was seen as a trouble maker and was soon dismissed from his post. Surprise, surprise, the abandonment of the hand washing policy saw mortality rates rise to their pre-Ignatian levels.

Frustrated, Semmelweis wrote increasingly furious letters and articles to the medical community, accusing them of cold-hearted murder. Accounts of his discovery were printed in journals, such as the *Lancet*. He repeated his successes, whilst working in hospitals in Budapest in the 1850s, and, in 1861, published his theory and statistical demonstrations in a book called *The Etiology, Concept and Prophylaxis of Childbed Fever*.

It was not well received.

Worse still, he became an obsessive on the subject at a time when he started to develop signs of the onset of what might have been Alzheimer's.

Even his wife thought he was verging on insanity. In 1865, he was lured into a mental asylum in Vienna. Realising he had been trapped, Semmelweis tried to make good his escape, but was detained, put in a straight-jacket and given a good hiding by the warders for good measure.

Two weeks later he died from his injuries, which had gone gangrenous.

It was only when Louis Pasteur was able to provide a theoretical explanation of the causal link between germs and disease that Semmelweis began to be regarded as the genius that he was and was able to claim his place as a pioneer of antiseptic policy.

35. Ludwig Boltzmann (1844 – 1906)

Science, in general, and physics, in particular, whilst fascinating, has always been a bit of a closed book to me. Thank goodness there have been cleverer people than I, who have made a significant contribution to the understanding of how the universe works. One such is Austrian-born Ludwig Boltzmann.

Take entropy. In a nutshell, it is a term used to measure the amount of energy in a system that cannot be used. Think of it as a thermodynamic waste product. The amount of entropy present in a system can be increased by a rise in temperature or the expansion of a gas or a solid turning into liquid.

I have always thought that tidying up was a bit of a waste of time and now I have the scientific evidence to back up my empirical observation. If I'm prevailed upon to tidy up a pile of clothing, have I contributed to a decrease in disorder and a corresponding reduction in entropy?

Not a bit of it.

You see, there are side effects to my attempt to restore order to my unruly pile of glad rags. I will be breathing, probably cursing, metabolising and warming my surroundings. When everything is totted up, the total disorder measured by entropy will have increased.

Boltzmann's contribution to the corpus of scientific knowledge was to apply statistical techniques to understanding the Second Law of Thermodynamics, first articulated by the French scientist, Sadi Carnot, in 1824, which held that the total entropy of an isolated system can only increase over time.

Boltzmann was an atomist and believed that these tricky little devils held the key to the understanding of entropy. By blending the Laws of Mechanics, as applied to the motion of atoms, with probability theory, he concluded that the Second Law of Thermodynamics was essentially a statistical law. The formula he derived to describe entropy in 1877 became the foundation of statistical mechanics.

Our hero didn't finish there.

Between 1880 and 1883, he continued to develop his statistical approach to explaining the mysteries of the universe and refined a theory to explain friction and diffusion in gases. In the late 1880s, following Hertz's discovery of electromagnetic waves, Boltzmann devised a number of experiments to demonstrate radio waves, and lectured on the subject.

Impressive as this all is, Boltzmann did not find favour with his colleagues.

Atomism, which is the bedrock of modern-day physics, was under attack at the time and Boltzmann's theory that entropy was irreversible was counter to prevailing thought at the time. After all, the equations of Newtonian mechanics are reversible over time and the great Poincaré had demonstrated that a mechanical system in a given state will always return to that state at some point.

One of Boltzmann's leading critics was Wilhelm Ostwald, who paid no heed to atoms, preferring to explain physical science purely in terms of energy conditions. Ostwald put the energist case against Boltzmann succinctly, recording that "the actual irreversibility of natural phenomena thus proves the existence of processes that cannot be described by mechanical equations, and with this the verdict on scientific materialism is settled".

Scientific discussions at the time were lively affairs, one contemporary describing a debate between Boltzmann and Ostwald as resembling "the battle of the bull with a supple fighter".

The constant criticism of his theories and the need to defend himself vigorously against all-comers wore Boltzmann down.

Whilst on holiday with his wife and daughter at the Bay of Duino near Trieste in 1906, he committed suicide by hanging himself.

Ironically, shortly after his death, discoveries in atomic physics such as Brownian Motion, the random movement of particles in a liquid or gas which can only be explained by statistical mechanics, reinforced the primacy of atomic theory and established Boltzmann's work as the cornerstone of modern-day physics.

36. Edouard-Leon Scott de Martinville (1817 – 1879)

It is always fascinating to hear yourself as others hear you.

Often it is quite a shock – do I really sound like that? The usual way in which we hear our voice as it really is by recording it on a tape recorder or a dictaphone and then playing it back. Of course, someone must have had the brain wave to capture the human voice and this is Leon Scott, to abbreviate the mouthful that is his name, comes in.

Scott was born and lived in Paris, a printer by trade. Perhaps unsurprisingly, he took some interest in the documents, journals, and books that he was printing. A particular speciality of his printing business was works of scientific interest and he was thus able to keep abreast with the latest developments.

Having seen the development of rudimentary cameras, which were able to capture images of the human form, he began to wonder whether a device could be built to record the human voice. Scott saw a particularly useful application in the ability to record a conversation verbatim, what we would now call stenography, and by 1849 had published several papers on the subject.

Proof-reading a physics textbook around 1853, he came across a series of drawings of the human auditory system and wondered whether it could be recreated mechanically. Naturally, he set out to see.

His design replaced the tympanum with an elastic membrane in the shape of a horn. For the ossicle, he devised a series of levers, which would move a stylus back and forth across a glass or paper surface, blackened by smoke from an oil lamp. The object of the exercise was to capture the sound of the human voice in a way that could be deciphered rather than played back.

Calling his machine a phonautograph, Scott sent a version of its design to the French Academy on March 25, 1857 and received a French patent for his troubles.

It is one thing coming up with an idea and another making some money out of it, the significant drawback to his design being that whilst it reproduced sound as a series of squiggles, it did not allow the recordist to play it back.

So, what sales Scott made were limited to the scientific community, principally to allow them to investigate the qualities and properties of sound. Laudable, for sure, but sales were insufficient to make a difference to his lifestyle and Scott saw out his days as a librarian and bookseller.

And there it may have rested.

But in 2008, a group of scientists from the Lawrence Berkeley National Laboratory got hold of one of Scott's phonoautographs and succeeded in converting the series of squiggles made on April 9, 1860 into a digital audio file. On playing it, they heard a twenty-second snatch of Scott singing, very slowly, part of *Au clair de la lune*, an audio recording pre-dating Thomas Edison's recording of the nursery rhyme, *Mary had a little lamb*, on August 12, 1877 by some seventeen years.

Edison received a patent for his phonogram later that year and Scott went to his grave convinced that the American had wrested some of the glory that was rightfully his. It is only now that his contribution to laying the foundations for recording the human voice has finally been appreciated.

37. Ole Johansen Winstrup (1782 – 1867)

As a twenty-five-year old, it must have been a particularly chastening experience to see your capital city set on fire by the British and what remained of your navy towed away.

The aftermath of the Battle of Copenhagen in 1807 caused Danish-born Ole Johansen Winstrup to exercise his grey cells to come up with an ingenious method of strengthening Copenhagen's sea defences.

In 1808, while he was a guardsman, he worked long and hard in his workshop to develop a model of what he called *Hvalfisken*, the Whale. And a pretty Heath-Robinsonish affair it was too. The idea behind the vessel was that the best way of surprising an enemy's fleet was to creep up on it from below. Essentially, The Whale was what we would now know as a submarine.

According to Winstrup's patent application, in preparation for an attack a diver would drill numerous holes into the submarine's hull and seal them with corks. When the vessel was in the requisite position, the diver would simply remove the corks, causing it to sink, forming a barrier against naval attack.

Presumably, once the diver had removed all the corks, he would swim to safety. Quite how many would have to be removed before the vessel became unstable was not made clear. Of course, there was the risk that the vessel would be detected and boarded by the curious enemy. Naturally, Winstrup had thought of that. "Should it happen that they send a ship's carpenter to examine the ship," he wrote, "then the harpoon shown in the model should be used."

Clearly satisfied that he had come up with a workable model and what would have been a game changer in the field of naval combat, Winstrup submitted his patent application to the Danish authorities. He even invited the Danish Crown Prince, later Frederik VI, to take a ride on the boat but the palace declined the kind offer.

It may have been this act of hubris on Winstrup's part that proved his undoing as the patent was refused "because of technical shortcomings" and the Whale was consigned to the scrap heap of history.

Bonkers as the idea of removing corks from holes to make a vessel sink may have been, careful inspection of Winstrup's plans would have revealed one revolutionary idea, the vessel used propellers.

Experimentation into mechanical power, principally driven by steam, was underway in various parts of the world in the early nineteenth century but Winstrup was ahead of the curve.

It was not until 1815 that Richard Trevithick had designed a steam-powered propeller and the late 1830s that John Ericsson came up with the two-screw propeller system for use on naval vessels.

The Danes had cavalierly thrown away a technological edge.

Although Winstrup gave up on the Whale, he did continue to blaze a technological trail.

In 1826, he built a two-horse-power steam engine, which was adopted by a Copenhagen brewery owned by Hans Bagger Momme. It was the first steam engine to be used in Denmark, built by a Dane. Winstrup built a few more steam engines and in 1827 he set up an iron foundry. He even operated a wind turbine.

However, Winstrup's more enduring claim to fame is coming up with the idea of using a propeller to drive a boat and not being able to convince the authorities to adopt the idea.

38. Ernest Duchene (1874 – 1912)

One of the few drugs to which I have an allergic reaction is penicillin, but there is no denying that its antibiotic properties have been a life saver over the last eighty years or so.

Alexander Fleming is credited with discovering the antibiotic properties of penicillin but this version of the history of the drug omits the contribution of Paris-born, Ernest Duchesne.

His tale is one of cheese, guinea pigs and misfortune.

Our hero enrolled into the military medical school in Lyon in 1894 and immediately became interested in the developing world of microbiology. His director of studies, Gabriel Roux, who was also responsible for monitoring the city's water supply, set him a challenge; why did tap water never go mouldy and yet mould grew easily in distilled water?

Using the mould that is used to make gorgonzola, stilton and other forms of my favourite blue cheeses,

Penicillium glaucum, Ernest put some in tap water and another batch in distilled water. The results were as Roux had observed. Even adding the typhoid bug and the e. coli bacteria to the culture produced the same results. There was clearly something in the tap water that was inimical to the mould.

The stroke of genius is to look at things in an unexpected way.

What fascinated Ernest was the mould and he chose to investigate whether there were any circumstances in which it would put up a fight against and even overcome hostile bacteria.

This is where the guinea pigs came in.

The first group he injected with strains of S. typhi or E.coli all died but the second group, who were also given a shot of Penicillium glaucum, were luckier; they became ill but recovered. A third batch, initially injected with P. glaucum, were even luckier; they proved immune to the diseases.

Somehow the drug created from the mould altered the virulence of the disease-carrying bacteria so that they were no longer deadly, pre-empting Fleming's discovery by some thirty-five years.

Duchesne wrote up his findings and submitted his thesis in 1897.

But then he started to be dogged by misfortune. Awarded his medical degree that year, he was commissioned as a second-class major of medicine in the French army. Marrying in 1901, his wife died two years later of tuberculosis. In 1904, he contracted some form of pulmonary infection, probably tuberculosis as well, and spent the final eight years of his life in and out of sanitoria.

Ironically, a few shots of the drug that he had discovered might well have saved his life.

Sometimes it is better to be a guinea pig.

Ernest never carried out any experiments after 1897, nor published any more papers and, whilst Fleming for discovering penicillin, and Chain and Florey for purifying it, were awarded the Nobel Prize in 1944, it was only in 1949, upon the rediscovery of his thesis, that his contribution to the discovery of penicillin was acknowledged.

And why was that?

Undoubtedly, Duchesne had conclusively demonstrated the anti-bacterial properties of moulds and even hinted that they may help to combat human infections. But P. glaucum would have been relatively ineffective and, possibly, lethal, if used to treat Homo sapiens. It produces something called patulin, much weaker than penicillin, which comes from P. notatum, and is only effective in much higher and toxic concentrations.

Duchesne had conducted the right experiment and had drawn the right conclusions. He had simply used the wrong mould.

Whether he would have moved on to P. notatum had he lived longer and had the opportunity to carry out more research will never be known.

Part Four

Patently unfair

PATENT NONSENSE

[THE PATENT SYSTEM] secured to the inventor, for a limited time, the exclusive use of his invention; and thereby added the fuel of interest to the fire of genius, in the discovery and production of new and useful things – Abraham Lincoln, *Lecture on Discoveries and Inventions,* 1858.

There is little doubt that the patent system has been a tremendous boon to the inventor but as with every system there are pitfalls, which the unwary need to avoid, not least of which is making sure that your patent is still valid and has not expired. There are companies out there, waiting to exploit an idea which has fallen out of patent.

Here are four cautionary tales that show what can happen when patents expire.

39. Mary Anderson (1866 – 1953)

It's an everyday scene. You jump into your car, notice the windscreen is a bit smeared, so you flick a switch and two mechanical arms, fixed to the exterior of your car, spring into action and clear it for you. When it is raining or snowing, the windscreen wipers are invaluable to help you see where you are going. Have you ever considered whose brainwave the wipers were?

This is where Alabama-born Mary Anderson comes in.

While in New York during the winter of 1902, Mary was travelling on a trolley car and it was sleeting. The stately progress of the vehicle was interrupted every now and again because the driver had to get out and clear the front window of the snow and ice that had accumulated there. Instead of fuming about the delay that this operation caused to her journey, Mary started wondering whether a type of blade could be produced to be operated by the driver from inside the trolley car, allowing him to clear the screen without having to stop and start the vehicle.

When she got back to Birmingham, Alabama, Mary's musings were sufficiently advanced that she was able to commit a rudimentary design to paper. She then wrote a description of how it might work and hired a local company to make a working model.

It was remarkably simple, consisting of a lever fixed to the inside of the vehicle, which controlled a rubber blade fixed to the exterior of the windscreen. By controlling the lever, the blade, which was counterweighted to ensure contact, would go back and forth across the windscreen, clearing it of any obstructions. The blade was detachable, "thus leaving nothing to mar the usual appearance of the car during fair weather".

On June 18, 1903, Mary submitted her application for a patent for what she quaintly described as a Window Cleaning Device. In the supporting documentation Anderson described how the wiper was to be operated by a detachable handle inside the vehicle.

On November 10, she was notified by the United States Patent Office that a patent had been granted, number 743,801, and that she had exclusive rights over her invention for seventeen years.

As we have already seen, inventing something is the easy part, making a commercial success of it is another kettle of fish altogether.

Mary started searching for commercial partners but, surprisingly, found no takers. Rather like an aspiring author seeking a publisher, she received rejections by the sack full. Perhaps the letter she received from the Montreal firm, Dinning and Eckstein, on June 20, 1905, was typical, stating that "we beg to acknowledge receipt of your recent favour with reference to the sale of your patent. In reply, we regret to state we do not consider it to be of such commercial value as would warrant our undertaking its sale".

So that was that and Mary seems to have abandoned her attempts to put her invention into production, concentrating on managing some flats she had built instead.

Her patent expired in 1920.

By that time, many more people owned cars and vehicle manufacturers were looking to enhance the specifications of their models. In 1922, Cadillac was the first to include windscreen wipers on all their models and soon they became standard equipment. The timing of their adoption was not coincidental, depriving Mary of cashing in on her simple but essential invention.

It was not until 2011 that Mary's contribution to vehicle safety was recognised by the Hall of Inventors.

40. Catherine Hettinger (1954 – present)

One of the challenges for an old fogey like me is to keep up with current trends. I'm told that a craze that swept through the playgrounds in 2017 was something called the fidget spinner. For those who are not in the know, it consists of a central circular pad, which the user holds, and two or three prongs, each holding a metal or ceramic bearing. The object of the exercise, if such a rudimentary process can be so described, is to rotate it between your fingers. Apparently, users enjoy a pleasant sensory experience. For those looking for more excitement, you can toss or twirl the spinner or transfer it between fingers.

What fun!

Proponents of the gadget claim that it helps relieve stress and is aimed at those children who suffer from ADHD, another of those conditions which seem to have sprung up since I was a child.

It certainly seems to appeal to those whose surfeit of energy is in inverse proportion to their concentration span. With the fidget spinner hailed as the toy of 2017, and flying off the shelves in their millions, you would think that the person who came up with the original concept would have unlocked the door to untold riches. But the cautionary tale of Florida-based, Catherine Hettinger, will dispel your illusions.

In the 1990s, Hettinger was suffering from myasthenia gravis, which causes your muscles to weaken. Desperate to keep her young daughter amused, she came up with a toy, which consisted of a circular device moulded from a single piece of plastic, which could be spun on the fingertip. In 1993, Hettinger applied for, and in 1997 was awarded, a patent for her device, described as a spinning toy. She toured around some of the arts and crafts fairs in Florida and sold enough to break even, improving on the design as she went along.

In search of her big break, our heroine approached toy manufacturing giant, Hasbro, who tested the design. Alas for Catherine, they decided not to put it into production. One of the problems with patents is that you need to renew them and this involves the periodic payment of a fee. Strapped for cash, Hettinger allowed the patent on her device to lapse in 2005.

In late 2016, eleven or so years after the patent lapsed, the Fidget Spinner began to make waves amongst the junior members of society and manufacturers of the toy started making lots of money. One of the ways that corporations can evade paying inventors their due is by making subtle changes to the design. Although the current versions of Fidget Spinners are spun using your fingertips, they rely on a completely different movement mechanism from Hettinger's prototype.

Worse still for Hettinger, even if she had renewed her patent, it would have expired in 2014, seventeen years after it had been granted.

This is the way that patents work, ostensibly giving an inventor enough time to capitalise on their genius without granting them a perpetual monopoly. You can't help thinking that the toy manufacturers waited until any vestige of patent right had disappeared before launching the Fidget Spinner commercially.

It is a moot point as to whether Hettinger would have had any entitlement to cash in. At the very least, she came up with the basic concept. Her story amply illustrates the inventor's lot.

41. George de Mestral (1907 – 1990)

One of the, admittedly minor, accomplishments of a child in my day was to be able to tie their shoelaces in a way that secured the shoes to their feet and avoided the risk of tripping on trailing laces. Nowadays, rather like the ability to wield a pen, this skill has an air of the recherché about it, thanks to the ubiquitous presence of Velcro.

And who do we have to thank for this time-saving convenience? Swiss-born electrical engineer and inventor, George de Mestrel, that's who.

Having mastered the art of tying my shoes, I used to love exploring the fields and country lanes near my home in rural Shropshire. Often, in ploughing through the undergrowth, my socks and clothing would be covered by those sticky and very adhesive burrs from the burdock plant. It gave me some amusement when I got home, picking them painstakingly off, but I never paid them much more attention.

George, on the other hand, had a much more inquisitive mind.

Returning from a hunting trip in 1941, he and his dog were covered in the burrs. He decided to look at one under a microscope and noticed that they were covered with thousands of hooks. If you brushed against a burr, these hooks attached themselves to you. In a lightbulb moment, George wondered whether he could replicate this irritating wonder of nature in fabric to produce a fastener, which would be easier to use than buttons and zips.

Realising that the hooks needed to attach themselves to something, he envisaged a piece of fabric with corresponding loops.

Et voilà, Velcro, a word coined from velvet and crochet, French for a hook.

George's problems, though, had only just begun. He approached half a dozen fabric manufacturers around Europe to see if he could interest them in his idea. They were sceptical as to whether the hooks could be mass-produced. Eventually, George found a manufacturer in Lyon who, by using a combination of nylon and cotton, was able to come up with something that passed muster. George patented his idea in 1955, borrowed $150,000, and set up a company to market Velcro.

Still, the ability to mass produce the fastener eluded him until, nearly twenty years after his brain wave, he came up with the design of a loom, which allowed him to cut the hooks at just the right angle to attach easily to the loops.

George was running out of money fast and the sales potential of Velcro was affected by the simple fact that it was unusual and didn't look good.

Velcro's fortunes started to take off when the material was adopted by NASA. Its adhesive properties worked well in the zero-gravity of space capsules and a picture of astronaut Buzz Aldrin showing off his watch with its Velcro band to Neil Armstrong saw interest in the fastener rocket. It was now cutting edge and Pierre Cardin, the French fashion designer, became obsessed with it.

The age of Velcro was born. Not that it did poor George much good.

By this time, he had sold his rights to the product to the Velcro companies and when he applied to update his patent, which had expired in 1978, he was rebuffed. Now it was out of patent, the fabric fastener was widely adopted.

Funny, that.

George had to content himself in the knowledge that Velcro was his brain-child. His experiences didn't dampen his inventive streak. He came up with a rather nifty asparagus peeler, always useful to have in the kitchen drawer, I feel.

George will be best known, though, for inventing Velcro and not profiting from its universal adoption, because his patent had lapsed.

42. William Austin Burt (1792 – 1858)

Such is the ubiquity of Global Positioning System (GPS) devices that it is almost a sign of eccentricity to be seen struggling with a map and peering at a compass. Of course, GPS is a recent innovation but two hundred years ago when there were still vast tracts of land to be explored and surveyed, at least from the invading white man's perspective, a magnetic compass was invaluable to navigate around terra incognita.

Magnetic compasses, though, were not infallible. In normal circumstances, the magnetised metal needle will settle down to show you where true north lies. From this simple observation, you can then work out which are the other major cardinal points of the compass and the direction that you need to follow. But the performance of the needle can be affected profoundly by the proximity of rocks containing a high metallic content. They can interfere with the compass needle, causing it to point in a direction other than true magnetic north.

William Burt is best known for developing and patenting the typographer, the first typewriter to hit the United States, but from 1833, he was working as a Deputy Surveyor, charged with surveying the then wide-open spaces of Michigan and Wisconsin. The latter state is particularly rich in iron ore deposits and Burt found that they were affecting the accuracy and consistency of his compass readings.

Realising the enormity of this problem and that a better, more accurate and reliable compass was needed,

Burt started to exercise his little grey cells to come up with an instrument that did not rely on magnetism.

His solution was a solar compass, made of brass with an attachment that allowed surveyors to determine true north by observing the sun.

It consisted of an arc for setting the land's latitude, another for establishing the declination of the sun, and a third for setting the time of day. All three arcs were placed on an upper plate, which was kept stationary when in use. The instrument's sights were placed on the lower plate, which could be clamped in any position to the upper plate.

Having created a working model in 1835, Burt submitted his design to the scrutiny of the Franklin Institute in Philadelphia. They awarded him a medal and twenty dollars in gold.

On February 25, 1836, Burt received a patent for his compass but continued to enhance its design over the next fifteen years. It was displayed at the Great Exhibition in London in 1851 and was awarded a prize.

Amongst his accomplishments as a surveyor were the discovery of the Marquette iron ore range in 1844 and establishing the northern portion of the border between Michigan and Wisconsin in 1847. The solar compass had more than proved its mettle and, for a century or more, became the standard piece of kit adopted by the General Land Office to be used in mineral-rich areas.

But trouble was looming. With his patent soon to expire, Burt went to Washington in 1850 to renew it.

The Land Committee, recognising how useful the compass was in surveying the immense tracts of the western United States, persuaded him not to renew the patent but rather petition Congress for compensation equivalent to the sum he may have otherwise generated from a reinvigorated patent. Burt followed their advice. After all, the Congress was full of honourable men.

What could possibly go wrong?

Whilst a payment of $300 was mooted, it never materialised. Now that the patent had expired, it left the way open for other instrument makers to supply what were known as Burt's solar compasses to surveyors.

The moral of this story is that if your patent is expiring, do not relinquish it for vague promises.

PATENT
DIFFICULTIES

A COUNTRY WITHOUT a patent office and good patent laws is just a crab, and can't travel any way but sideways and backwards – Mark Twain, *A Connecticut Yankee in King Arthur's Court*, 1889.

Mark Twain may have shared Abraham Lincoln's enthusiasm for patents and their supporting legal framework, but not all inventors would have agreed. The proof of the pudding is in the eating, as the proverb says. Just because you have a piece of paper saying that the design and concept is yours, it doesn't mean you can sit back and enjoy the fruits of your labour.

Here are six stories, recounting the difficulties that inventors had in enforcing their patent rights. The process was exhausting and took its toll on their health and mental wellbeing.

43. Edwin Howard Armstrong (1890 – 1954)

Before the arrival of Digital Audio Broadcasting (DAB), what revolutionised the quality of radio signals was the switch from Amplitude Modulation (AM) to Frequency Modulation (FM). Gone for the most part were those irritating hisses from extraneous noises, which affected and spoiled many an AM radio programme. The battle to get FM adopted, though, was long and hard and eventually did for its inventor, Edwin Armstrong.

In the mid-1920s, Armstrong began researching into ways to eliminate the static that bedevilled AM radio, initially by modifying the characteristics of existing AM transmissions. Finding little success, in 1928, he turned his attentions to investigating the use of frequency modulation transmissions.

Working away in a laboratory in Columbia's grandiosely named Philosophy Hall, he developed what is now known as wide-band FM. It had significant advantages over the previously developed narrow-band transmissions, principally because it allowed more information to be carried and was more robust and resistant to interference. He was granted five US patents on December 26, 1933 on the basic features of his new system.

That was the easy part.

Armstrong had a standing arrangement with RCA to give them first refusal on his patents and, whilst they were impressed, they were focusing their investment strategy on the nascent television technology and so declined the opportunity.

Undaunted, Armstrong decided to finance his own development and form relationships with smaller players in the radio industry, like Zenith and General Electric.

In June 1936, Armstrong demonstrated his new system to the US Federal Communications Commission (FCC), in front of an audience of 500 engineers, by playing a jazz record through conventional AM radio frequencies and then through FM. A contemporary noted that "if the audience… had shut their eyes they would have believed the jazz band was in the same room. There were no extraneous sounds".

But Armstrong hit major problems.

A switch-over to the ultra-high frequency system would mean scrapping all the existing broadcasting equipment and domestic radios, an expense which for America, just emerging from the Great Depression, was unpalatable.

Then, when interest in FM did grow amongst some of the radio stations, construction restrictions that were put in place during the Second World War limited its growth. To prevent interference between radio stations that were early adopters of FM and the mainstream AM stations, the FCC reallocated the FM band, to 88 to 108 Mhz, which meant that the existing FM equipment and receivers had to be scrapped. Armstrong saw the dead hand of RCA behind the attempts to frustrate the adoption of FM.

Armstrong's fourth problem was his long legal battle with RCA.

In 1940, they had offered him $1 million for a non-exclusive, royalty-free licence to use his FM patents but Armstrong turned them down. This prompted RCA to conduct their own research into FM and develop what they claimed to be a system which didn't infringe Armstrong's patents. Worse still, RCA encouraged other companies to stop paying royalties to Armstrong.

In 1948, our hero sued RCA and NBC, claiming patent infringement and that they "had deliberately set out to oppose and impair the value" of his invention. The case dragged on, depleting Armstrong's finances, worsened when his primary patents expired in late 1950.

It all got too much for him and, during the night of January 31, 1954, Armstrong jumped to his death from a window of his flat on the thirteenth floor of River House in New York City.

His wife, Marion, pursued the case against RCA and reached an out of court settlement of around $1m. It was not until the 1960s that FM started to become established in America, although NASA adopted Armstrong's system for communications between Houston and the Apollo astronauts.

It was a long and arduous battle and, ultimately, it proved too much for Armstrong.

44. Philo T Farnsworth (1906 – 1971)

After a hard day at work many of us spend more time than we would care to admit, slumped somnolently in front of a glowing rectangular box transmitting what passes for entertainment these days.

Yes, the television.

A man who played a crucial part in the development of the television set we have come to know and love was Utah-born scientist, Philo T Farnsworth. He had to fight a protracted legal battle in order to assert his rights and it is only recently that his contribution to the development of the TV has been truly recognised.

This is his story.

Philo, who had already shown his mettle as a child by winning a national contest for inventing a tamper-proof lock, was an avid reader of science magazines. He became interested in the concept of television and quickly deduced that the mechanical systems that were being suggested would be too slow to scan and assemble the many images required to put on a moving picture show.

In a chemistry lesson at school, he sketched out an idea for a vacuum tube that would revolutionise the TV, although no one realised it at the time. By the age of sixteen, he had worked out the basic outlines of a functioning electronic television.

In 1926, Philo raised some money to fund his work, $6,000 from private investors and $25,000 from Crocker First National Bank of San Francisco, and, on September 7, 1927, made his first successful electronic television transmission, filing for a patent that year.

Continuing to work on and perfect the equipment, Farnsworth gave his first demonstration to the press in September 1928.

But trouble was just around the corner. His backers, keen to capitalise on their investment, began talks with RCA. RCA sent their head of TV, Vladimir Zworykin, to review Farnsworth's work.

Zworykin was by no means an impartial assessor, after all, he was working on similar ideas for the American corporate, and concluded that whilst his receiver, the kinescope, was superior, Farnsworth's video camera tube, which dissected images and was essentially what he had sketched out in his science lesson a few years earlier, was the bee's knees.

To buy him out, RCA offered Farnsworth $100,000, an offer he rejected.

The 1930s saw Farnsworth embroiled in legal battles with RCA, who claimed that his inventions were in violation of a patent filed earlier than his by Zworykin. The resources of RCA funded a series of actions, appeals and counter-appeals and it was not until 1939 that they agreed to pay Farnsworth $1m for his patents. The Second World War put a stop to TV production and, by the time peace returned, Philo's patents had expired in any case.

The decade of legal battles had taken its toll on Farnsworth's health, causing him to have a nervous breakdown in the late 1930s. However, in 1947 his company Farnsworth Television produced its first TV set.

The company, though, was unable to compete with the giants of the industry, particularly RCA, got into financial difficulties, and was taken over by IT&T in 1949. Farnsworth was retained as vice president of research but the battle for primacy in the TV market was lost.

Worse was to follow.

He moved back to Utah to continue research on technologies such as radar, infra-red telescopes and nuclear fusion but his company, Philo T Farnsworth Association, went bankrupt in 1970. Philo then took to drink and died of pneumonia in Salt Lake City on March 11, 1971.

It was only through the efforts of his wife, Elma aka "Pem", that Farnsworth's part in the development of TV has been belatedly recognised. In 1999, he was included in *Time* magazine's Time 100: The Most Important People of the Century and was inducted, posthumously of course, into the Television Academy of Fame in 2013.

But it was a long hard struggle to gain the recognition he was due.

45. Gary Kildall (1942 – 1994)

The early days of computing were a bit like the Wild West, where fortunes were won and lost and where the unwary were ripped off.

What has been particularly liberating for the ordinary users of PCs has been the development of operating language and protocols, which are intuitive and easy to use. You would think that the person who made all this possible would have been assured of fame and riches.

Think again, as I recount the salutary tale of Gary Kildall.

Kildall started working for Intel in 1974, hired to develop programming tools for their 4004 micro-processor. He was a bit of a whizz at coding software, turning out code that would work, albeit not necessarily the final polished article. He proposed and developed a high-level language for the 8008 and 8080 models, which would allow the user to issue quasi-English commands to the chip rather than relying on binary code.

Intel developed the now unfortunately named ISIS, which was the world's first floppy disk-based system, and

Kildall developed the operating system, CP/M. When Intel decided not to make the PC available in the commercial market, Kildall obtained their consent to market and sell a version of PC/M. Kildall set up Intergalactic Digital Research, there is nothing like having overarching ambitions, and soon PC/M was the operating system of choice in the nascent personal computing world.

But success was not assured.

In 1980, IBM was on the hunt for an operating system for their soon to be released PC and naturally contacted our hero. Legend has it that Kildall preferred to fly his aeroplane rather than meet the suits from IBM but the reality is that whilst he was late for the meeting, the sticking point was that the hardware giant was only prepared to pay him a one-off fee of $200,000 for using his operating system.

DOS – heard of that? – now enters our story.

It was developed by Seattle Computer Products and purchased by a company called Microsoft, owned by one Bill Gates – heard of him? This operating system took the best features of the CP/M operating system but tweaked it ever so slightly to make it incompatible with Kildall's offering. Gates, a smarter business man than Kildall, sold the rights to IBM for a paltry $50,000, reckoning, rightly as it turned out, that he could make much more by licensing it out to other computer manufacturers, the archetypal sprat to catch a mackerel.

Kildall threatened to sue but never did and was forced to develop another operating system, DR-DOS, to compete with what was his own system.

It never gained much traction and, in 1991, Novell bought what was left of his company. Gates had pretty much the whole of the market sewn up and when Windows replaced DOS, his place in history was secured. Kildall was relegated to a footnote, if that.

Worse was to follow.

In 1994, Kildall walked into a bar in Monterey, wearing a leather jacket with Harley-Davidson badges. There was a group of bikers in the bar and an altercation broke out. Kildall was pushed, fell to the floor, and died from head-related injuries. The coroner reported that the death was suspicious but no one was held to account.

A sad end for the man who developed the first PC operating system but, perhaps, typical of the ill luck that dogged him.

46. John Fitch (1743 – 1798)

Lady luck plays a large part in someone's success. If you are cursed with bad luck, then it is even harder to reap the rewards that your invention merits. A case in point is the story of the American, John Fitch.

Born in Connecticut, Fitch was a bit of a jack-of-all-trades in his youth, turning his hand to farm work, clock making, silversmithing, cartography, and fighting in the Continental Army during the American Revolution. After his discharge, he explored the Ohio River valley and was captured by a group of Native Americans, who turned him over to the British. Eventually, he was released but perhaps it was this experience that caused him to ponder whether there was a method of propelling river craft more quickly than simple muscle power.

Fitch's idea was to deploy the new-fangled steam powered engines that were beginning to make their mark in Britain. Engines, he surmised, would enable boats to move up and down rivers independently of concerns such as tides and weather.

Unfortunately for Fitch, one of the consequences of American independence was that the British refused to share their new technology with their erstwhile colonists and so he had to start from scratch, deploying the services of a clockmaker, Henry Voight, to build an engine.

By this time, he had persuaded various state legislatures to grant him a 14-year monopoly for steamboat traffic on their inland waterways, a concession that enabled him to raise investment from prominent Pennsylvanian businessmen.

The first public trial of Fitch's steamboat, appropriately named *Perseverance*, took place on the Delaware river on August 22, 1787, in front of assembled dignitaries. Although successful and drawing fulsome praise, no additional funding was forthcoming.

Undaunted, Fitch and Voight built a more substantial vessel, sixty feet long with a steam engine powering a number of oars positioned in the stern, which paddled rather like a duck. During the summer of 1790, Fitch carried up to thirty passengers a time on journeys between Philadelphia and Burlington, travelling in total over 1,500 miles at speeds averaging six miles per hour but getting up to a racy eight miles per hour at times.

Just as importantly, Fitch claimed they could travel upwards of 500 miles without any mechanical mishap.

Although Fitch was awarded a patent on August 26, 1791 for his steamboat, after a ferocious battle with James Rumsey, who had also invented a steam-powered vessel, it did not grant him a monopoly. It just protected his design. This caused many of Fitch's investors to jump ship and our hero was left high and dry.

Desperate for funding, he went to France but arrived at the height of the revolutionary turmoil that was the Reign of Terror. The monied classes had more pressing concerns on their collars. When a fund-raising trip to England also drew a blank, Fitch returned to the States.

Misfortune continued to dog him.

He moved to Kentucky, where he had bought some land in the 1780s, hoping to sell some of his newly acquired acres to finance the building of a steamboat to ply the Ohio or Mississippi rivers. But he found them occupied by settlers, necessitating a protracted legal battle to evict them.

Fitch continued working on steam engine concepts. Found in his attic after his death was a design for something described as "the prototype of a land-operating steam engine" meant to operate on tracks. His train preceded that of Richard Trevithick's, built in 1802 and recognised as the daddy of the steam locomotive.

Alas, Fitch fell into depression, drank heavily, and committed suicide in 1798, allowing Robert Fulton, with better financial backing, to steam in and make his dream of steam-powered boats a reality.

47. Peter M Roberts (1945 – present)

Here's a cautionary tale about employee suggestion schemes, involving socket wrenches and Peter M Roberts.

Socket wrenches have been around since medieval times and were used, for example, to wind up clocks. The first ratcheting socket wrench with interchangeable sockets was invented by an American, J.J Richardson, who filed a patent for his tool on June 16, 1863.

Although immensely useful, interchangeable socket wrenches were cumbersome as the operative had to stop what they were doing and use both hands to change the socket.

Roberts' light bulb moment was to make the operation much slicker by developing a simple, quick-release device, which allowed the user to change sockets quickly and easily with one hand.

He even developed a prototype.

At this time the eighteen-year-old Roberts worked for the retail chain store, Sears, in Gardner, Massachusetts, but all the development was done in his own time, not his employers'. So pleased was Roberts with what he had produced that he was about to hire a lawyer and file a patent, when he made a fatal mistake.

He mentioned what he had done to his boss.

The boss, in what was possibly one of the worst pieces of mentoring advice in modern history, suggested that Roberts enter his invention into the employee suggestion scheme. After all, Sears were selling around a million wrenches a year and would be bound to be interested.

This Roberts did on May 7, 1964 with a note stating that a patent application was pending. He made the even more calamitous mistake of surrendering the only prototype in existence.

Having received this gift horse, Sears proceeded to put the device through a series of tests and received the thumbs up from wrench operatives. By this time, Sears had closed the store in Gardner in which Roberts was working. As he was out of work, he went back to Tennessee to live with his parents. Sears compounded their ill-treatment of him by claiming that there was no commercial value in the device, although, by way of a sop, they gave him $10,000 for the patent.

The reality was somewhat different. Market research had convinced Sears that they were on to a winner and the product was launched in October 1964. Within a year Sears had sold twenty-six million of the wrenches, realising a profit of some $44 million.

By 1982 they had sold some thirty-seven million.

The only contact Roberts had from Sears during this time was a phone call asking for the identity of his patent lawyer, whom they promptly hired to protect their interests!

Realising the enormity of his mistake, Roberts started to bombard Sears with law suits, claiming that they had defrauded him.

The path to justice is long, tortuous, and expensive and it was not until 1976 that Roberts succeeded in getting a US Federal jury to agree with him and award him $1 million in damages, a paltry amount considering the success of the product but for someone on their uppers welcome indeed.

Sears were not finished with Roberts and decided to appeal the decision, taking the case all the way up to the Supreme Court, although they eventually lost.

The litigation continued and Roberts was able to up the damages awarded to him to $5 million. But even then, the dispute dragged on and it was not until 1989, some twenty-five years after the wrench had been invented, that the case was settled, Roberts walking away with $8.9m. This was enough for him to establish Link Tools which, surprise, surprise, manufactured quick-release ratchets, sockets and accessories.

Roberts was successful in the end but it had been a long hard struggle.

48. Charles Francis Jenkins (1867 – 1934)

Whether we like it or not, popular entertainment was transformed in the early twentieth century by the development of television and cinematography. Someone who could justifiably claim to be at the birth of both media is Charles Francis Jenkins. Much good it did him.

Born to a Quaker family, who moved, when Jenkins was just two, to farm in Fountain City, Indiana, as a boy he was forever tinkering with machinery and soon proved to be a dab hand at fixing broken-down implements. He also showed an inventive streak, developing a jack to lift wagons so their axles could be greased.

Like many a youth, Jenkins could not resist the lure of the city and, at the age of nineteen, moved to Washington DC, working as a stenographer in the early incarnation of the US Coast Guard. Although he had left his country roots behind, Charles could not shake off his inquisitiveness.

By 1890, Jenkins began working on what he described as a "motion picture projecting box" and called a Phantoscope. By the spring of 1894, he was sufficiently satisfied with his progress that he wrote to his parents that he was coming back to Indiana to show them his latest invention, instructing them to assemble a crowd of relatives and interested bystanders at his cousin's jewellery store in Richmond on June 6.

The gadget was packed up and sent to Richmond, Jenkins following on, completing the 700-mile journey by bicycle!

After some technical issues, according to the *Richmond Telegraph*, "there began a spluttering sound as the machine kicked into life and out of the lens shot light onto the wall and a girl clad in garments more picturesque than protective stepped lively. She did not seem bashful thus displayed, while those in the audience were taken aback." The shameful hussy was Annabelle, a vaudeville favourite.

The audience, after recovering from this assault on their sensibilities, went behind the screen to check that there had been no sleight of hand. Not only was this the earliest documented performance of moving pictures to an audience but, astonishingly, it was in colour, as each frame had been stained or coloured by hand. Moreover, it used reeled film and an electric light to project the images.

In the winter of 1894, Jenkins was introduced to Thomas Armat, who was looking for investment opportunities. Jenkins was strapped for cash and, by March 1895, they concluded an agreement by which Armat would "finance and promote the invention" of Jenkins.

The duo patented the Phantoscope on August 28, 1895 and gave a public demonstration of their device at the Cotton States Exposition in Atlanta in the autumn of 1895. A modified Phantoscope was patented on July 20, 1897 but relations between the two began to deteriorate. Jenkins eventually sold his interest in the projector to Armat, who then sold the rights to Thomas Edison and the rest is history.

But Jenkins wasn't finished as an inventor.

He developed a spiral-wound cardboard container, the design is still used today, a car with an engine in the front rather than under the driver (in 1898), an early version of a sightseeing bus (in 1901), an automatic starter for cars (1911), and an improved internal combustion engine (in 1912).

In an article entitled *Motion Pictures by Wireless – Wonderful possibilities of Motion Picture Progress* which appeared in the *Movie Picture News* of September 27, 1913, Jenkins announced that he had developed a mechanism which enabled him to view distant scenes by radio or what we would nowadays know as television. Notwithstanding his enthusiasm, it took him another ten years before he was able to transmit a picture, of President Harding, from Washington to Philadelphia but, by 1925, he was beaming moving pictures.

Granted a patent (US No 1,544,156 for Transmitting Pictures over Wireless) on June 30, 1925, Jenkins established the first commercially licensed TV station in America,

W3XK, which made its first transmission on July 2, 1928 from Washington. In 1929, it was broadcasting five nights a week.

It initially broadcast silhouettes but later moved on to transmitting black and white programmes. Jenkins' company even produced the equipment that early adopters would have to use to receive the pictures.

But timing is everything. Selling expensive and, essentially, novelty equipment and services as America was plunging into the depths of the Depression was not a smart move. Jenkins' company was declared bankrupt in 1931, opening up a space for RCA to exploit. And exploit it, they did.

CAVEAT VENDITOR

A HALLUCINATION IS a fact, not an error; what is erroneous is making a judgement based upon it – Bertrand Russell, *On the nature of Acquaintance: Neutral Monism,* 1914.

We have all heard of caveat emptor, the Latin motto that encourages the would-be purchaser to look carefully at what it is they propose to buy. There is, however, the other side of the coin. Caveat venditor encourages the prospective seller to conduct due diligence on their putative customer. They may not be all they seem to be.

Laudable as the patent system is, it has made the patent a tradable asset. There are individuals and companies around who are prepared to buy up patents. For the impecunious inventor there is undoubtedly an attraction to cash a patent in, trading uncertain and immeasurable future revenues for the certainty of immediate cash.

Or, at least, that is what they thought.

We conclude our survey of the perils that can beset an inventor by looking at a couple of individuals who came unstuck when they sought to cash in their chips.

49. Joseph Hansom (1803 – 1882)

I was rereading one of Arthur Conan Doyle's Sherlock Holmes stories the other day. The protagonist rushed to the scene of the crime in a Hansom cab, the principal form of taxi at the time. It set me thinking about who designed the carriage and this led me to the unfortunate character that was Joseph Hansom, whose ingenuity and ill-fortune earns him a place in our survey of the inventor's lot.

Working as an estate manager at Caldecote Hall, near Nuneaton, Hansom came up with a revolutionary design for a safety cab. It could hold two passengers with the driver seated at the back, communication between the two parties being effected through a trapdoor in the roof.

Its principal advantage over contemporary rivals was that it had a low centre of gravity, large wheels and a lower cab, and suspended axle, which meant it was much more stable when cornering. Being light and capable of being drawn by only one horse, making it cheaper for the cabbie to operate, it was faster and more manoeuvrable than many of its rivals.

Hansom applied for a patent on December 23, 1834 and the first Hansom cab travelled down the Coventry Road in Hinckley in 1835.

The design was a great success and Hansoms soon replaced the more expensive to run four-wheeled Hackney carriages as the vehicle of choice for hire. In its heyday, there were up to 7,500 Hansoms plying their trade in London and they were to be seen in other major cities in the UK, as well as Paris, Berlin, St Petersburg, and New York.

The last London Hansom driver handed his licence in as recently as 1947.

Although others, notably John Chapman, made modifications to the design, mainly to improve passenger comfort, Hansom's design stood the test of time. As the holder of the patent, you would expect Joseph to have received a handsome reward for his ingenuity.

Alas, he didn't.

He sold his patent for the cab to a company for the sum of £10,000. The company immediately got into financial difficulties and reneged on the payment, leaving Hansom without a penny.

Throughout his life Hansom was dogged by ill-luck.

Starting out as an architect, he designed over two hundred buildings including Plymouth Cathedral. Hansom and his partner, Edward Welch, overcame stiff opposition to win the commission to design and build Birmingham Town Hall in 1831. It is a beautiful building with tall pillars and a Roman feel about it, but costs soon spiralled out of control and, as the architects had stood surety for the builders, the edifice brought Hansom's company crashing down into bankruptcy.

In 1843, Hansom, together with Alfred Bartholomew, started an architectural journal called the *Builder*, which is still going today, although it was renamed *Building* in 1966. Aimed at architects, builders and workmen it found a profitable niche but, as you might expect, Hansom didn't share in the rewards. He had to relinquish his control over the journal because of lack of capital.

Whilst his name was immortalised in the cab that he designed, there is a blue plaque in his memory outside one of his former residences, 27 Sumner Place in South Kensington, he didn't receive the money that his ingenuity deserved.

It must have been particularly galling for him to summon a cab.

50. Walter Hunt (1796 – 1859)

Born in Martinsburg, in upper New York state, Walter Hunt trained as a stonemason but ended up working in a local flax mill. He had an inquisitive mind and an inventive streak, going on to become a serial inventor. Initially, he began to potter around to see if he could develop a more efficient form of flax spinner.

Naturally, Walter could and sufficiently encouraged by his model, he applied for, and was granted, a patent in 1826. Recognising that his machine was a game changer, he wanted to build a business around his invention, but there was one problem. He didn't have the financial resources to bring his plans to fruition.

The solution was to treat his patent as a commodity and sell it to the highest bidder. This seemed to be Walter's modus operandi throughout his career.

And a prolific career it was too.

Among his many inventions, the list is too exhaustive for this vignette, were a coach alarm system, which allowed a coachman to warn pedestrians of oncoming horses, a nail-making machine, a ship which broke up ice, a knife sharpener, a rope-making machine, and a street sweeper. Where they were patented, Walter soon sold them on.

Another of Walter's brainwaves was to develop a repeating rifle and cartridge system, the design of which would be used by Smith and Wesson. Naturally, Walter saw little financial reward for this innovation.

Some of Walter's inventions were off the wall, or not, in the case of what was known as an "antipodean apparatus". Despite its odd name, it was a pair of shoes, which allowed the wearer to walk up walls and ceilings. It went down a storm amongst circus performers. It continued to sell and be used until well into the 1930s, but despite its apparent success, Hunt was on his uppers.

In what must be an early example of inventor's remorse, wishing that he could put the genie he had released back into the bottle, Walter made a significant breakthrough in the development of the sewing machine. In 1833, he came up with what was the first workable sewing machine. He was concerned that if the machine took off it would damage the employment prospects of seamstresses and so, true to form, sold the rights to a businessman.

The businessman struggled to manufacture the machine commercially and gave up, crucially omitting to patent the design. That seemed to be the end of the story until, in 1846, Elias Howe was awarded a patent for his sewing machine.

Howe was disputatious and launched a series of lawsuits against other sewing machine manufacturers to protect and assert his patent rights. This alerted Walter to the fact that Howe's design was not dissimilar to the one he developed thirteen years earlier. After a legal battle, Hunt was recognised as the inventor, but the absence of a patent meant that Howe got to keep the intellectual property rights to the machine.

Isaac Singer now enters our story.

His iconic sewing machine, the prototype of the machine we know today, incorporated elements from Hunt's and Howe's design. Howe took Singer to court for Patent Infringement. In his defence, Singer claimed that Howe had ripped off Hunt's design. The absence of a patent on Hunt's machine counted against Singer, who had to pay Howe substantial damages.

As a by-product of this case, Singer eventually agreed, in 1858, to pay Walter $50,000 for incorporating elements of his design in his machine but then fate intervened. Walter died of pneumonia in 1859, before he had received a cent from Singer.

Perhaps Walter's greatest contribution to modern life was his invention of the humble safety pin. It is a pin with a spring mechanism and a clasp which fastens the pin to whatever it is to be attached to and prevents the user from pricking their finger.

The design is so simple and effective that it is hard to envisage how it can be improved upon. The story goes that fretting over a $15 debt, Walter was fiddling with a bit of wire. In a flash, the idea of a covered pin came to him and within a few hours, had completed his design. Although he patented the design, naturally, he sold it on for somewhere between $100 and $400, a fraction of what he could have earned from it.

But, then, that is the inventor's lot.

CONCLUDING THOUGHTS

MANY THINGS HAPPEN between the cup and the lip – Richard Burton, *The Anatomy of Melancholy,* 1621.

Our survey of the lot of the inventor has revealed that there are plenty of obstacles to overcome before an inventor can enjoy the fruits, both financial and reputational, of their ingenuity and labour. Some can be ascribed to poor choice or the pressures and prejudices exerted by the society in which they lived or the predatory behaviour of others. To navigate through these choppy waters and to ensure that the opportunity is not missed, the inventor needs more than their fair share of luck and for the fickle finger of fate to beckon kindly in their direction.

Whilst there remains a spark of ingenuity, creativity, and inquisitiveness residing in the human brain, there will be those amongst us who will seek to further our understanding of the world we live in and to make our daily existence more comfortable. After all, that is one of the traits that marks us out as human.

The lot of an inventor is not always an easy one and luck and the fickle finger of fate will continue to determine who succeeds and who languishes, forgotten in the mists of time.

At least I hope I have played my small part in righting some injustices and allowing some of the lesser known inventors to take their place amongst those who have improved our lot. To all of them, we owe our heartfelt thanks.

APPENDIX ONE

MARTIN FONE'S ILLUSTRIOUS HALL OF FAME

O VER THE COURSE of two books, I have shone the light on inventors who have come unstuck in some way, in an attempt to bring their stories back to life so that we can enjoy their genius and learn from their errors and misfortunes.

My initial concept was to create a Hall of Fame so that they would not be forgotten. It would be remiss of me, therefore, if I did not, by way of this appendix, remind you of the fearless one hundred whose fates reveal the interplay between luck and success.

I have used the initials, TFF for *The Fickle Finger* and FCB for *Fifty Clever Bastards* to denote which book you can find their stories in.

Valerian	Abakovsky	FCB
Abu Nasr	al-Jawari	FCB
Mary	Anderson	TFF
Edwin Howard	Armstrong	TFF
Harvey	Ball	FCB
James Miranda Stuart	Barry	FCB
Trevor	Baylis	FCB
Lszlo	Biro	FCB
Alexander	Bogdanov	FCB
Ludwig	Boltzmann	TFF
Karlheinz	Brandenburg	FCB
John Romulus	Brinkley	FCB
William Austin	Bullock	FCB
Robert	Bunsen	FCB
William Austin	Burt	TFF
Cowper Phipps	Coles	FCB
Martha	Coston	TFF
Karl	Drais	FCB
Charles Francis	Drew	TFF
Ernest	Duchene	TFF
Geoffrey	Dummer	TFF
Albert	Einstein	TFF
Douglas	Engelbart	FCB
Philo T	Farnsworth	TFF
Guiseppe	Fiechi	FCB
John	Fitch	TFF
Stephen	Foster	TFF

Rosalind	Franklin	TFF
Sieur	Freminet	FCB
Luigi	Galvani	FCB
George de	Garrett	FCB
James	Goodfellow	FCB
Charles Francis	Goodyear	FCB
Wilson	Greatbatch	FCB
Rasmus Malling	Hansen	TFF
Joseph	Hansom	TFF
William Austin	Harvey	FCB
Catherine	Hettinger	TFF
Elias	Howe	FCB
Horace	Hunley	FCB
Walter	Hunt	TFF
Daisuke	Inoue	TFF
Charles Francis	Jenkins	TFF
Thomas L	Jennings	TFF
Gary	Kildall	TFF
Marie	Killick	FCB
Ron	Klein	FCB
Margaret	Knight	TFF
Otto	Lilienthal	FCB
Abraham	Lincoln	TFF
Charles Francis	Lindbergh	TFF
Hans	Lippershey	TFF
Elizabeth J	Magie	TFF
Edouard-Leon Scott	Martinville	TFF
Elijah	McCoy	TFF
Lise	Meitner	TFF
John Joseph	Merlin	TFF
George de	Mestral	TFF

Antonio	Meucci	FCB
Thomas	Midgley	FCB
Benjamin T	Montgomery	TFF
Garrett	Morgan	TFF
Tom	Ogle	FCB
Thomas	Paine	TFF
Denis	Papin	FCB
Cecilia	Payne	TFF
Richard	Pearse	TFF
Arthur Paul	Pedrick	FCB
Anthony	Pratt	FCB
Louis le	Prince	FCB
Robert	Recorde	FCB
Peter M	Roberts	TFF
Angela Ruiz	Robles	TFF
Wilhelm	Rontgen	TFF
Sylvester	Roper	FCB
Jean-Francois	Rozier	FCB
Jonas	Salk	FCB
Carl Wilhelm	Scheele	FCB
Ignaz	Semmelweis	TFF
Walter	Shaw	FCB
Jerry	Siegel	TFF
Louis	Slotin	TFF
David M	Smith	TFF
Henry	Smolinski	FCB
Sabin Arnold von	Sochosky	TFF
Percy	Spencer	FCB
Nettie	Stevens	TFF
Hugh Edwin	Strickland	FCB
Joseph	Swan	FCB

Lewis	Temple	TFF
Konstantin	Tsiolkovsky	TFF
Max	Valier	FCB
Edward	Vernon	FCB
Ruth	Wakefield	TFF
John	Walker	FCB
Dietrich Nikolaus	Winkel	TFF
Henry	Winstanley	FCB
The	Winstons	FCB
Ole Johansen	Winstrup	TFF
The	Xerox Corporation	FCB

More about the author

If you enjoyed this book, check out Martin Fone's other books:

Fifty Clever Bastards – a study of luck and success or a simply a feast of schadenfreude, featuring fifty inventors who came a cropper; either they were killed by their inventions, were ripped off or simply gave their inventions away for the good of mankind.

Fifty Curious Questions – an attempt to answer some of those irritating questions that life throws up along the way. The selection is idiosyncratic and is designed to show the lengths scientists have gone to and the quantum leaps in logic they have deployed to push out the frontiers of human knowledge. The book was a Category Finalist in the prestigious 2018 Eric Hoffer Book Award.

Fifty Scams and Hoaxes – a light-hearted investigation into the murky world of financial skulduggery, medical quackery and ingenious hoaxing. This book was a Category Finalist in the Independent Author Network Book of the Year Award, 2019.